THE FATE OF
GREENLAND

THE FATE OF
GREENLAND

LESSONS FROM ABRUPT CLIMATE CHANGE

PHILIP CONKLING, RICHARD ALLEY,
WALLACE BROECKER, AND GEORGE DENTON

PHOTOGRAPHS BY GARY COMER

The MIT Press
Cambridge, Massachusetts
London, England

For information about special
quantity discounts, please email
special_sales@mitpress.mit.edu

This book was set in Chaparral
and Gotham by The MIT Press.
Printed and bound in Canada.

Library of Congress Cataloging-
in-Publication Data

The fate of Greenland : lessons
from abrupt climate change /
Philip Conkling . . . [et al.].

p. cm.

Includes bibliographical references and index.

ISBN 978-0-262-01564-6
(hardcover : alk. paper)
1. Climatic changes—Environmental aspects—Greenland.
2. Global warming—Greenland.
3. Greenland—Environmental
conditions. I. Conkling, Philip W.

QC903.2.G83F38 2011

551.69982—dc22

                    2010040698

10 9 8 7 6 5 4 3 2 1

To the Memory of
Gary Comer
Who took us to the far ends of the Arctic
and left a living legacy of a new generation
of climate scientists

CONTENTS

In August 2001, Gary Comer, the transoceanic sailor who founded the Lands' End direct mail clothing empire and who had been fascinated with the Arctic since childhood, successfully completed a voyage from Greenland through the Northwest Passage. For centuries mariners had tried to navigate through ice-choked channels of the Northwest Passage that connect the Atlantic and Pacific oceans at the top of the world. Even the names of these channels conjure up the images of the men and their sponsors who tried and mostly failed to find a route through this treacherous Northwest Passage—Franklin Bay, Peel Sound, Prince Regent Inlet, Coronation Gulf, Amundsen Gulf, and the Beaufort Sea.

Comer's was the first private voyage through this legendary passage completed in a single season without the services of a government icebreaker—and the fastest in history, completing the crossing in sixteen days and eight hours. That accomplishment changed his life and set the stage for his decision to fund important scientific research that has altered our understanding of the global extent of abrupt climate change.

Although Comer was thrilled by the unexpected success of the expedition, he recognized that the rapid melting of the sea ice that enabled him to complete his voyage presaged massive environmental changes to the Arctic. Shortly after completing his historic journey, Comer sold Lands' End and was free from a lifetime of business responsibilities. A modest, self-effacing man who had gone to work to support himself after high school, Comer began visiting scientists around the United States and asking them to instruct him on how the changes he had witnessed in the Arctic would affect the rest of the world, since he sensed that the region might be a distant early-warning area for the effects of global warming.

I had been aboard various expeditions with Gary Comer to remote parts of the northern oceans, informally assigned the role of expedition naturalist aboard his 152-foot vessel, *Turmoil*. After Comer's successful crossing of the Northwest Passage in 2001, I suggested he contact Wallace "Wally" Broecker, an oceanographer at Columbia University's Lamont-Doherty Earth Observatory. From background

reading, I knew that Broecker had spent a half-century studying how oceans and thus climates have swung back and forth between warm and cold phases in the earth's geologic past and that he was the recipient of numerous awards, including the Crafoord Prize, the "Nobel Prize" for scientific fields lacking a Nobel. When they met at Broecker's lab, he told Comer of the recent scientific discoveries that showed how abruptly climate had changed in the past and described his concern for the unpredictable effects that future climate change represents to life on the planet as we know it.

In the absence of any significant leadership in Washington at the time to fund a coordinated abrupt climate change program, Comer decided to fund a major effort himself and asked Broecker to help organize an interdisciplinary program to recruit talented young scientists to improve the understanding of this field of research.

Both iconoclasts in their own areas of endeavor, Comer and Broecker hit it off immediately. In Broecker, Comer found a world-class intellect, a scientist with blunt opinions and a respected antibureaucratic nature that reflected Comer's own deepest inclinations. Under Broecker's guidance, Comer began investing tens of millions of dollars to work with a network of the most distinguished scientific mentors in the world to identify promising young PhD candidates and newly minted post-doctoral students to track changes in the earth's ocean and atmospheric systems across the globe. Broecker quickly recruited George Denton, a geologist from the University of Maine, and Richard Alley, a glaciologist from Penn State, to help coordinate the Comer Fellows Program in Abrupt Climate Change Research.

During the Arctic summers of 2002, 2003, 2005, and 2006 Comer, Broecker, Denton, Alley, and their students organized a series of research expeditions to the Arctic, and in particular to Greenland, while Comer also battled cancer, which ultimately claimed him in 2006. Comer invited scientists to travel aboard the *Turmoil*, to identify sites for further field research. We traveled widely throughout vast areas of the Arctic, including several voyages to western, southern, and eastern Greenland. With *Turmoil's* cruising range of 10,000 miles, accompanied by either an amphibious float plane or a helicopter that could land on *Turmoil's* afterdeck, the scientists had undreamed-of access to nearly any site they wanted to visit from the western Canadian Arctic to eastern Greenland. Comer, a highly talented photographer in his own right, documented these expeditions.

Figure P.1
Gary Comer in his amphibious floatplane over Scoresby Sound, 2003. (Photo by Philip Conkling)

Figure P.2
Gary Comer just prior to his successful crossing of the Northwest Passage. (Photo by Philip Conkling)

Figure P.3
On the stern of *Turmoil*, 2005 (left to right); George Denton, Richard Alley, Philip Conkling, Gary Comer and Wallace Broecker. (Photo by Philip Walsh)

By 2003, under the banner of the Comer Fellowship Program, Broecker, Denton, and Alley had established a network of 25 senior scientists to focus on the global dynamics of abrupt climate change. With the commitment of significant multiyear funding, Broecker, Denton, and Alley fashioned a scientific program to blend both model-based and paleoclimatic approaches to understanding climate change, with one important caveat: they focus on the study of changes in the earth's climate that occur not slowly over tens of thousands of years, but over periods of decades or years.

What Gary Comer's support has accomplished in the most profound sense is to have helped scientists who study abrupt climate change to understand more of the risks we face and encourage a prudent response. Like a venture capitalist, Comer found the people with big ideas and small means, relying on Broecker, Denton, and Alley to find the brightest young scientists at the beginning of their careers to focus on the big questions rather than on fund-raising at a critical point in their careers. Because Comer was the "venture capitalist," these efforts are not isolated from the broader scientific community, but rather are the framework on which the broader scientific community is building. The venture capitalist invests in the hope of greater riches. Comer had already been wildly successful in the business world, and then invested for a much bigger payoff, to change how we understand the world.

Philip Conkling, Editor

ACKNOWLEDGMENTS

The authors wish to acknowledge the many people who helped make this book possible. Wendy Strothman, of the Strothman Agency, initially suggested that a book based on the scientific adventures of our group in Greenland could find a market and introduced us to Clay Morgan of MIT Press, who was enthusiastic from the start.

Stephanie Comer, Guy Comer, and Bill Schleicher of the Comer Science and Educational Foundation generously supported the publication of photographs from Gary Comer's archive. Alison McKinzie expertly produced the image files from the Comer archive and Bridget Leavitt of the Island Institute prepared the additional image files for this volume. Patty Catanzaro of the Lamont Laboratory of Earth Sciences at Columbia University helped prepared charts, maps, and other figures.

Philip Walsh, Beverly Walsh, and the late Telford Allen graciously and professionally supported all of the field research expeditions aboard Gary Comer's vessel, *M/V Turmoil* when the authors conceived and discussed this book project. Scott Travers helped support the annual Comer Fellows meetings at Leeward Farm, where many of the ideas discussed in this volume were first presented. Peter Quesada, a companion on many of the expeditions offered helpful comments on several early drafts of the manuscript.

In addition to Clay Morgan, the team at MIT Press was consistently helpful and diligent, including our editor, Michael Sims, Laura Callen, Susan Clark, Virginia Crossman, and designer Emily Gutheinz. At the Island Institute, Gillian Garrett-Reed helped correct the page proofs, while Kathy Allen provided invaluable administrative support.

The authors also wish to acknowledge the support of their families who stoically put up with missed anniversaries and birthdays during the years we were away on field expeditions, and especially our respective wives, Paige, Cindy, Elizabeth, and Marna.

Ultimately, this book rests on the excellent work of the 74 Comer Fellows and their 18 mentors who continue to produce invaluable scientific information that will help add to the sum of our knowledge in understanding the past reality and future challenge of abrupt climate change.

LESSONS FROM ABRUPT CLIMATE CHANGE

Greenland's Prospects

Greenland is the world's largest island, 90 percent of which is covered by ice. Greenland is also one of the remote wonders of the world. Greenland's ice sheet—the largest outside Antarctica—stretches almost 1,000 miles from north to south and is 600 miles east to west. The view from a small plane out over the endless desolation of snow and ice in the high summer light is an experience in the incredible whiteness of being. Scattered here and there over the ice sheet surface are little ponds of azure-colored melt water that present a striking contrast to the all-encompassing whiteness below.

When the artist Rockwell Kent went to Greenland to live in 1931, he described Greenland as buried under a vast ice sheet with "a narrow strip of mountainous land between the ice cap and the sea. It is a stark, bare, treeless land with naked rock predominating everywhere. Few countries of the world, or none, that are inhabited by man have less to offer man for ease, security of life, and happiness." But that view is changing—and changing rapidly.

We would expect that Greenland, as a remote part of the earth at high latitude, would be an obvious place to observe and study the effects of climate changes over time. And in fact, historical research has clearly documented that

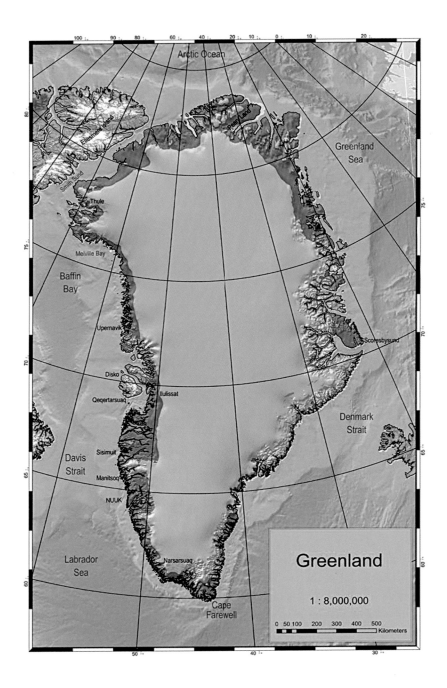

Arctic Ocean

Ellesmere Island

Peary Land

Greenland Sea

Smith Sund

Thule

Melville Bay

Baffin Bay

Upernavik

Disko

Ilulissat

Qeqertarsuaq

Scoresbysund

Denmark Strait

Sisimuit

Davis Strait

Manitsoq

NUUK

Labrador Sea

Narsarsuaq

Cape Farewell

Greenland

1 : 8,000,000

0 50 100   200   300   400   500
Kilometers

Greenland has twice experienced dramatic and rapid shifts in climate during the last thousand years of its history—both for better and for worse.

The dates of the Norse settlement and occupation of Greenland—from the late 980s to 1410—neatly circumscribe most of the era that climatologists call the Medieval Warm Period, which was an interval of notably warm temperatures across northern Europe. The years of comparative warmth during the ninth through twelfth centuries helped enable long Norse voyages first to the Shetlands, then to the Faroe Islands, then to Iceland and Greenland and ultimately to Labrador and Newfoundland. Contemporaneous records from England show that the last two hundred years of this period, between 1100 and 1300, were especially warm in Europe, and that agricultural crops from southern Europe, such as barley, oats, and wheat, spread far into northern Europe for the first time.

During one of our research voyages, we visited Erik the Red's farm at Brattahlid ("Steep Slope"), in the region of protected southern fjords, which inspired Erik to name this new land "Greenland." Like Erik, we were struck by the vivid color of the surrounding verdant pastures, where a new generation of Greenlanders was using tractors on the hillsides to cut hay for their large flocks of sheep. Erik fortuitously immigrated to these new lands at the beginning of the Medieval Warm Period when he and his fellow colonists established two Norse outposts—the Eastern Settlement around Brattahlid and the Western Settlement, in a fjord 400 miles north on Greenland's west coast. These settlements survived for almost half a millennium before they mysteriously disappeared at the beginning of another abrupt climate change.

Most scholars have concluded that a catastrophic abandonment occurred in the Western Settlement shortly after 1349, the year word reached the Norse Bishop at Gardar in southern Greenland that the Western Settlement was in great trouble. It took another two years for the bishop to get a boat there, whereupon the bishop's men found only sheep and cattle wandering in the hills. Although in the more populous Eastern Settlement, the Norse continued to hunt, farm, and trade, for at least a half-century after the abandonment of the Western Settlement, life there appears to have been increasingly tenuous. The year 1367 provided the last record of a Royal Ship reaching Greenland bringing the last bishop of Greenland, who presided for a decade until his death in 1378. In the 1380s, Icelandic annals mention only four or five trading expeditions between Iceland and Greenland as sea ice increasingly closed off contact.

Figure I.1
Map of Greenland.

The horrific dangers of navigating through the ice in the Greenland Sea during storms and persistent fogs even in the best of summers had undoubtedly begun to strangle trade and communication. Successful summer expeditions to the northern hunting grounds in the spring for food would have been increasingly difficult to mount without any support from the abandoned Western Settlement. Prolonged cold winters and short rainy summers made raising livestock increasingly marginal. The church, a drain on resources during the best of times, exercised less and less control, particularly when Norway failed to replace Greenland's last bishop after 1378. Starvation must have been a major preoccupation. In perilous circumstances, isolated farms may have been tempting targets for outsiders—Inuits or pirates—to attack and plunder.

Perhaps we will never know what happened to the Greenland Norse at the onset of the advancing cold period climatologists now call the Little Ice Age. What we do know is that at the time of the beginning of the Little Ice Age in the fourteenth century, a civilization of perhaps as many as 4,000 to 5,000 Greenland Norse at its peak had survived for 450 years before disappearing almost without a trace. As scientific evidence becomes more precise in dating changes in the climatic conditions in Greenland and in the North Atlantic, we will gain a greater appreciation of how one civilization at the center of a rapidly changing climate responded to their predicament and suffered their ultimate fate. But whatever discoveries await us, we know that Greenlanders were at the epicenter of two changing climate regimes in the North Atlantic, which coincide with establishment and abandonment of Greenland's Norse colonies.

Greenland Today

The research voyages that Gary Comer sponsored all around Greenland's coast and over its ice sheet documented a rapidly changing climate at every location we visited. At Ilulissat in 2001, we watched the flow of icebergs calving off the Sermeq Kujalleq glacier, as the native Greenlanders call it (the Danish call it Jakobshavn Isbrae), where the flow was constant and breathtaking. The glacier's sheer size was difficult to comprehend because the scale of the surrounding landscape was massive. We used the ship's sonar to determine that the glacier's

Figure I.2
A meltwater lake on the surface of the Greenland ice sheet. Ice flow over bedrock bumps creates local low spots on the surface where water pools. These lakes sometimes wedge open crevasses, draining suddenly and spectacularly through the ice to the bottom and then out to the edge of the ice. (Photo by Gary Comer)

Figure I.3
Icebergs from the great Sermeq Kujalleq, or Jakobshavn Glacier, which drains roughly 7 percent of the Greenland ice sheet, producing immense numbers of bergs, probably including the one that sank the Titanic. (Photo by Gary Comer)

bottom lay in 1,200 feet of water and with a sextant measured its height at another 600 feet in the air overhead. During the course of several days, we heard loud cracks shattering the stillness every five or ten minutes as another massive wall of ice split off the face of Sermeq Kujalleq and crashed into Disko Bay to become part of a flotilla of approximately 10 percent of the icebergs set adrift in the North Atlantic from this one glacier alone.

More recently scientists who have implanted GPS sensors in the glacier "upstream" have documented that its speed has dramatically increased in recent years to as much as 30 meters per day, double the rate of a few decades ago. Scientists, including Richard Alley, hypothesize that the Sermeq Kujalleq's speed can be traced to large pools of water that melt in the summer on top of the vast Greenland ice sheet, filling crevasses where water pressure can drill a shaft, or *moulin*, from the surface of the ice cap two miles down through the ice sheet. The cascading sheets of water thaw the bottom of the glacier and unglue it from its rock bed where it is otherwise anchored and begin to lubricate its sleigh ride to the sea.

In Scoresby Sound on the east coast of Greenland in 2003, the evidence of the warming climate was also apparent everywhere we stopped. After crossing the Greenland Sea from Iceland, we anchored off a headland in a cove near the entrance. George Denton, the expedition's field geologist, said, "I'll show you what global warming looks like." He pointed out the ship's window to a distinctly visible line on the rock wall 20 to 30 feet above the fjord's waterline separating a facade of black bedrock from white rock underneath. Denton pointed out that the blackish rocks higher upslope were coated by dark lichen while the whitish rocks underneath had no lichen. Denton said that the white rocks had still been scoured by a glacier during the Little Ice Age that scientists date as lasting from approximately 1350 until 1880. Since the end of that cold period, there had not been time for the lichen to colonize the white rocks that had been scraped clean. But no ice was in sight—it had completely melted during the past century. In a later flight over Liverpool Land Denton said, "These glaciers are all dying—they're basically all ablation [melting] zone. They'll be gone entirely in twenty years."

Two years later, in 2005, we approached the face of the Qaqat Glacier in southern Greenland. Navigating with electronic charts updated in 2001, we ap-

Figure I.4
A debris-laden iceberg in Scoresby Sound. Glaciers carry material that falls on them, or that they pick up from beneath them or bulldoze in front. This material is eventually left in ridges called moraines, or spread across the landscape or dropped from icebergs to the sea floor. (Photo by Gary Comer)

proached the edge of the ice field, with the electronic monitor in the pilothouse showing us our position. According to the chart plotter that interfaced with Global Positioning Satellites (GPS) to display our position in real time, we could plainly see that the edge of ice field had receded, although the position of the shoreline and other surrounding geographic features appeared surprisingly accurate. When Philip Walsh, captain of the *Turmoil*, took a chart scanned from Danish naval charts with 1982 soundings and overlaid it on our current position, the effect was startling: Qaqat glacier had receded 2.3 miles back up the valley in 18 years.

One of the more lasting impressions of the fate of Greenland occurred after a night in a lodge in Ilulissat in 2003, while we waited for a part for the floatplane. The lodge was full of tourists, attracted in part to numerous boutiques on the main floor as well as the deck overlooking the harbor where a profusion of grand icebergs slowly drifted by on the West Greenland current. A well-appointed, middle-aged European couple came out of one of the boutiques. The man was completely decked out in seal fur—seal pants, a seal vest, a seal sports coat, and a seal cap. It was difficult not to gape; obviously he did not expect to meet a Greenpeace or PETA activist in Ilulissat. From at least some of the natives' and tourists' points of view, the rapidly melting glaciers of Greenland are good news because the regular and photogenic displays of icebergs calving will bring ever more well-heeled tourists to witness the spectacle.

But what will Greenland's warming climate mean to the rest of the world?

The Clues to the Future Are in the Past

This volume hopes to capture the intense excitement of the past decade of research that continues to improve our scientific understanding of what Greenland's ice cap, glaciers, and seas are telling us about how climate changes, and how through linkages in the natural system those changes have spread across the globe. The discoveries of the scientists who have been leaders in this field, especially Wallace Broecker at Columbia University, George Denton at the University of Maine, and Richard Alley at Penn State, along with the contributions from more than a hundred of their colleagues who have been working around the

Figure I.5
Large iceberg calved from
Daugaard–Jensen Glacier,
Northwest Fjord in Scoresby
Sound. (Photo by Philip
Conkling)

world for much of the past decade, have brought the remote island of Greenland to the center of the understanding over how abruptly climate has changed in the past.

Essentially there are two approaches to understanding climate change. The first, called the paleoclimatic approach, unravels climates of the past in order to determine what physical processes may have triggered those changes. The other approach involves building complex computer models based on equations that attempt to quantify what scientists know about the earth's terrestrial, oceanic, and atmospheric systems. If you want to understand how expensive components will behave in a complex system, scientists and engineers build carefully scaled models to test them. Before building a supertanker, say, that will carry millions of barrels of oil across the oceans of the world, you might find it prudent to build a scale model and test it in a large wave tank simulating ocean conditions, for example, during storms. It is not the same thing as knowing how a tanker will fare in a storm, but it helps predict the stresses on the vessel. Similarly, before redesigning the economy of the world to reconfigure our reliance on carbon-based fossil fuels, scientists build models to provide information for policy makers who ask what will happen to the world if we double or triple the amount of carbon dioxide in the atmosphere.

Modern climate modeling was built on the foundation of weather-forecasting a half century ago as atmospheric scientists used the principles of physics to simulate how water vapor condenses, forms into clouds, and rains on our picnics, and how warmer temperatures stir the air into winds and eddies to form such features as the jet stream or tropical hurricanes. Eventually atmospheric models were "coupled" to ocean models based on oceanographic data, and later modeling of the terrestrial biosphere was added. Compiling all of these coupled models on giant supercomputers now enables scientists to understand the immense complexity of the planet as a system of systems. Applying them to climate conditions during past epochs enables scientists to test these models to see if they correctly simulate the results that scientists have reconstructed. If the models accurately simulate what happened in the past, scientists gain confidence in them.

The combination of field research with modeling "experiments" then enables climate scientists to ask important questions about the how climates of the future might affect life on earth.

Figure I.6
George Denton and Gary Comer on glacial outwash at Sydcap, Scoresby Sound. (Photo by Philip Conkling)

Figure I.7
Entrance to Scoresby Sound after crossing the Greenland Sea from Iceland. (Photo by Gary Comer)

Richard Alley's Climate Switch

In the opening chapter of this book, we introduce the theory of ice ages, as developed by Richard Alley, one of the country's foremost glaciologists, and explain how increases and decreases in carbon dioxide in the atmosphere have affected climates in the past. In the second chapter, we describe the role Alley and others played in the discovery of abrupt climate change.

Beginning in the late 1980s and for most of the first half of the 1990s, the National Science Foundation in the United States and a European consortium of major scientific academies sponsored two drilling projects on the Greenland ice cap. From the earlier drilling projects, scientists had learned that energy from the summer's sun changes snow to produce annual snowfall layers, which can be counted like tree rings. They also knew that anything in the air that falls with the snow—such as forest-fire smoke and volcanic ash from great eruptions—would also appear in the annual layers.

Trapped bubbles of old air contain carbon dioxide, which can be studied for any correlation with the advance and retreat of ice sheets. The bubbles also contain "swamp-gas" methane, which can tell scientists how widespread wetlands were. Scientists were thus able to sample air bubbles trapped in individual layers of ice to measure CO_2, methane, and other "greenhouse gases" locked inside to "see" the composition of the atmosphere as the climate changed. After drilling down two miles and extracting the cores, scientists from the United States and Europe could produce a year-by-year climate history extending back for over 100,000 years.

Beginning with the Greenland ice drilling projects of the 1990s, sophisticated new techniques were deployed for the first time to measure the history of temperature and snowfall locked in individual layers of ice, which could then be connected to specific years in history and then to distant prehistory. Perhaps no one fully anticipated the wealth of results that these ice cores would reveal. The drilling projects produced startling results that clearly demonstrated how past climates were racked by abrupt and almost inconceivably dramatic temperature changes that reorganized the climate across most of the earth.

Alley spent four seasons on the Greenland ice cap as part of the Greenland Ice Sheet Project (GISP) that ultimately drilled and retrieved a two-mile-long ice core with its distinct annual and seasonal layers back to the surface for analysis.

Figure I.8
Statue of Leif overlooking his father's, Erik the Red's, farm Brattahlid. (Photo by Philip Conkling)

During that time Alley worked in an ice tunnel painstakingly studying changes in past snowfall and temperature from the drill cores. Alley vividly recalls his excitement as he began counting the years of the core back from the present: "This snow fell the year I was born, that snow the year Lincoln spoke at Gettysburg."

The most important results of the ice core projects began to appear in the scientific literature by 1993, most notably a paper in which Richard Alley collaborated, "The Flickering Switch of Late Pleistocene Climate Change," published in the scholarly journal *Nature*. But the paper was largely ignored in the popular press and even today most people regard climate change as a gradual, if inexorable, process that plays out generally in incremental changes over time. However, that perception is beginning to change. Alley used a metaphor to describe his "flickering switch," comparing it to "flipping a kayak." Lean a little in a kayak to watch the aquatic life beneath you, and the boat tips a little. Lean a little too far and you cross a threshold, flipping the boat. The climate system, too, has thresholds, and changes in carbon dioxide or in Earth's orbit or in other things have sometimes "rocked the boat" until it flipped over into a new mode that could be either dramatically warmer or dramatically colder.

In 2000 Alley sharpened his concern in his book, *The Two Mile Time Machine*. He wrote, "The ice age took 10,000 years to end, but much of that change happened in less than 10 years." With abruptly warming temperatures, scientists could plainly see in the ice core record that summer snowfall doubled, with most of the change occurring in a single year. Subsequently, chemical measurements by ice core project chief scientist Paul Mayewski, now at the University of Maine, showed that huge changes happened over a few years to a very few decades across much of the rest of the world, including warming of about 18 degrees Fahrenheit in Greenland in about ten years.

Figure I.9
Greenland's ice cap is nourished by storms that deliver large amounts of snow to the top of the ice sheet. Much of the ice sheet is surrounded by spectacular mountains, especially on the east and south. (Photo by Gary Comer)

Broecker's Ocean Conveyor

Wallace Broecker, or "Wally" as everyone calls him, is the eminence grise of climate change research. In 1960 he published his first scientific article on the subject of using radiocarbon dating of deep ocean cores that serve as time capsules for unraveling past ocean circulation. In 1975, Broecker published a seminal

article in *Science*, "Climate Change: Are We on the Brink of a Pronounced Global Warming?" thus bringing the phrase global warming into both the scientific and popular lexicon. Twelve years later Broecker followed up with an article in Britain's leading scientific journal *Nature*, "Unpleasant Surprises in the Greenhouse?" and has likened Earth's climate to "an angry beast that we are poking with sticks."

Broecker was the first scientist to pick up on the then almost unnoticed research of Helmut Heinrich, who presented strong evidence that past ice sheets exhibited large instabilities, dumping debris-laden icebergs into the ocean in great surges. Heinrich's hypothesis, backed by additional research findings of Broecker and others, began to focus attention on the modern ice sheets. Richard Alley first described how a crevasse opens and how streams of melt water plunge through the ice to lubricate the bed, how temperature is the main control on the size of the ice, how warming can rapidly affect the flow, and how enigmatic events such as earthquakes may be harbingers of change.

Broecker's most notable intellectual leap is his ocean-conveyor hypothesis, which suggests that the ocean currents of the world are tied together in a vast globe-girdling conveyor belt, with the critical link driven by the sinking of cold salty water, most dramatically in the seas surrounding Greenland. The dense sinking water snakes through the abyssal depths of the North and South Atlantic ocean basins, all the way to the antipodean margins of Antarctica's continental slope. There the water races around the continent in the great Antarctic circumpolar current before splitting into two branches. One branch loops north into the Indian Ocean, while another surges into the vast Pacific Ocean, ultimately rising to the surface at the far northern edges of the Pacific basin. There this great upwelling conveyor loops around as a warm surface current to complete an enormous interconnected multiocean crossing that takes a thousand years. Broecker's ocean conveyor hypothesis is a breathtakingly simple conceptual model based on an enormous volume of detailed scientific research from many different fields.

Broecker has also proposed a compelling explanation for what happens to the world's climate when the ocean conveyor shuts down. If you want to stop a conveyor belt at an airport, you quickly realize it is most vulnerable to disruption at points where a loop turns back down on itself. If the seas around Greenland were to become fresher from the influx of a melting ice sheet or the collapse of an ice dam that releases large amounts of water into the North Atlantic, the ocean conveyor could slow or stop.

Broecker has calculated how much heat the conveyor carries northward from the tropics: it is comparable to nearly a third as much energy as the sunlight that falls on the entire North Atlantic. Without the Atlantic's warm surface ocean currents and the Gulf Stream carrying tropical heat to the northern parts of Europe, wintertime temperatures would plummet. Most people are surprised to learn that today in England, for example, roses grow further north than the latitude where Canadians encounter polar bears. More worrying than whether London, which is at the same latitude at St. John's Newfoundland, might experience snow squalls in the summer is the question of what would happen if the ocean conveyor were to shut down. Broecker has suggested that in the past a shutdown of the conveyor was accompanied by drying of the great Asian monsoon, which now waters the crops of billions of people across the most populous continent from India to China.

In chapter 6, we describe how abrupt climate shifts in a tightly coupled climate system have propagated across different regions of the earth. Sudden and severe droughts, including ones in the U.S. southwest, have repeatedly occurred in a warming world. Studies of stalagmites from China and Brazil and ocean sediments in a Florida lake show how the Asian monsoons and Caribbean rainfall have turned on and off in their respective regions in near lockstep with the retreat of glacial moraines across several mountain ranges of the world in the space of a few decades. All of these findings in the field have been combined with targeted modeling to synthesize the results, which have been among the Comer fellows' major contributions.

Denton's Seasonality and Conveyor Wobbles Hypotheses

If Wally Broecker is one of science's big thinkers, you might think of George Denton, University of Maine's eminent geologist, as his down-to-earth reality check. Broecker describes himself as having "a mind that leaps around a lot." Part of the reason for Broecker's long friendship with Denton is that he has helped Broecker "ground truth" or field-test some of his most important ideas.

George Denton specializes in tracking glaciers, and he has mapped glacial features, most notably moraines, on all seven of the earth's continents during a lifetime of field research. Denton's contributions to the scientific literature on ice ages have shown how even a little atmospheric cooling causes ice to advance down mountain valleys, since at such times glaciers do not retreat in the summer, while they advance further each winter. Repeated global cooling has spread ice sheets across vast expanses of the world. When warming drives the ice back, the piles of debris left behind as moraines become forensic chalk lines showing where the ice was when it died. By carefully dating those moraines in any of several ways—from counting rings in trees growing on top, to carbon dating bits of shells or wood at the edge of a moraine—Denton has carefully reconstructed the long history of glacier warming and cooling across all seven continents.

In this volume, we describe Denton's two major hypotheses that have contributed directly to the understanding of the dynamics driving abrupt climate change. During a visit to Scoresby Sound in Eastern Greenland with Gary Comer, Richard Alley, and Philip Conkling in 2003, Denton immediately recognized from his deep knowledge of glacial landscapes that winter temperatures during a period of a rapid climate reversal called the Younger Dryas (about 11,500 years ago) had to have been much colder than previously thought. During that same period, indications were that summer temperatures in eastern Greenland were generally moderate. The only way such an anomalous situation could have occurred was if sea ice had covered much of the North Atlantic, diverting warm Gulf Stream waters in the ocean conveyor farther south.

In chapter 5, we present Denton's proposal that the Little Ice Age that drove the Norse civilization out of Greenland resulted from a wobble in the current of Broecker's North Atlantic conveyor that changed the amount of summer ice in the region, dooming the Norse. To understand how Earth's most recent rapid climate change affected a hardy civilization of Norse who had lived in Greenland for half a millennium, it is useful to compare paleoclimatic scientific research with historical records and archeological research from this period of Norse occupation of Greenland.

In this regard, Richard Alley's observations from examining the Greenland ice cores corroborate that this warm period also occurred in Greenland, where it is evident in the isotopic composition of the ice and trapped gases, and still exists as a slightly warm zone in the ice.

Figure I.10
Sermeq Kujalleq, or Jakobshavn Glacier, near Illulissat, produces spectacularly large icebergs. Fresh-water streams from under the melting glacier and the icebergs it produces bring nutrient-rich water to the surface to help support a local halibut fishery. (Photo by Gary Comer)

The Fate of Greenland's Ice Sheet

At the end of this volume, we turn back to the question of the fate of Greenland's ice sheet. Richard Alley has asked himself why Greenland's ice sheet still exists when nearby Baffin Island's ice sheet at the same latitude and altitude is gone. Alley points out that Greenland's ice cap is nourished by storm tracks that deliver large amounts of snow to the top of the ice sheet where it does not melt. Indeed, recent changes are causing the ice cap to grow slightly higher at its center, while melting at its edges. As a result the entire Greenland ice sheet is developing a steeper slope aspect, causing scientists to ask whether it is becoming less stable. At Ilulissat (Jakobshavn), the most famous iceberg-producing area in the world, the huge expanse of the ice sheet is forced through a small opening between the headlands, producing spectacularly huge icebergs when releases occur, although not all of Greenland has such channels.

Alley and other scientists have calculated that if the Greenland ice sheet melts, sea level would rise seven meters—or about 24 feet—worldwide. In contrast, if the West Antarctic ice sheet melted, there would cause a five-meter—16-foot—sea level rise; the huge East Antarctic ice sheet is believed to be too cold to change dramatically, although the coastal regions of similar size to the West Antarctic ice sheet might also be vulnerable to global warming. Already, rising sea levels have begun to seep into the footings in lower Manhattan at the site of the World Trade Tower reconstruction. If the Greenland ice sheet melts, most all of lower Manhattan would either need to be protected by dikes or would be underwater and Florida's coast would recede nearly to Orlando. If warming also destroys the East Antarctic ice sheet and causes an additional 15-foot sea level rise worldwide, most of the state would disappear.

The big news is that worldwide temperature spikes occur when the global climate is changing modes. At such times climates change with such stunning speed that adjustments, even by the rich and technologically sophisticated nations of the world, may barely mitigate the effects. And for poor countries, the losses will likely be catastrophic.

Most of the scientists in the Comer program, along with thousands more who have worked on Intergovernmental Panel on Climate Change (IPCC) panels, believe that if we are on the verge of a change of climate modes from one of relative stability to instability, then continued loading of carbon dioxide into the atmosphere only increases the uncertainty and the instability with which we will have to contend.

We have become the biggest force in the climate system, pushing it faster than the imperceptible variations in the sun's energy reaching the earth that have contributed to past abrupt climate changes. We can have a huge effect on the *rate of change* the future will bring. Given the levels of uncertainty we face and the risk that catastrophes await us, wisdom dictates caution.

You might think of exercising caution as analogous to purchasing insurance. Spending relatively small amounts of each nation's GNP to reduce carbon emissions does not guarantee that all droughts, floods, hurricanes, and crop failures will be avoided. Similarly, buying homeowners insurance, installing a lightning rod, and cleaning up oily rags in the corner of a garage do not guarantee that your house will not burn down. You cannot know whether your house will burn or not, but what prudent person or nation would want to risk its health, wealth, and future without insurance? We do not yet have the scientific sophistication to know the likelihood that an abrupt climate change will occur at any specific time in the future, but we now know that abrupt climate changes have repeatedly occurred in the past and that we are taking an enormous risk with every additional ton of carbon dioxide that enters Earth's atmosphere and oceans.

Climate change research has elegantly demonstrated that the best we can hope for is gradual warming in response to rising carbon dioxide from our burning of fossil fuels. Too often in the past, however, the climate has staggered drunkenly, a phenomenon that is deeply worrisome because faster and unexpected changes are harder to deal with and responding to them is much more costly. The research described in this volume has established unequivocally that these sudden stumbles are real.

This book presents the evidence that Greenland has experienced two abrupt climate changes in historic times and scores of abrupt climatic changes in recent geologic history as documented in Greenland's ice records. Greenland appears to be poised at the edge of another rapid climate change, which in the past has propagated climate changes across both hemispheres. Therefore, it is in all of our interests to pay attention to Greenland, because in the fate of Greenland lie clues to the fate of the world.

Greenland, in other words, matters.

1

MYSTERY OF THE ICE AGES

The Theory of Ice Ages

Tourists love to get near glaciers, but most people prefer to live elsewhere, and for good reasons. Strong, cold winds often drain down-valley from the icy glaciers. The soils just beyond the ice are usually rocky and poor, and often there is no soil at all, just bare rock or gravel. A hungry farmer looks elsewhere for a new row to hoe.

Centuries ago, the people who lived near glaciers, whether high in European valleys or elsewhere, knew that the glaciers grew and shrank with the changing climate. At least some of those people realized that the stony soils, bare rocks, and other features being formed by the glaciers were nearly identical to features down the valley, and those people correctly inferred that the glaciers had once been bigger.

But in eighteenth-century Europe, many university scientists did not seriously consider the possibility that they lived on the deposits of glaciers. The evidence was everywhere, but the idea seemed monstrously strange to many, and only a few were engaged in asking why the landscape looked as it did.

A glacier moving down a valley may bulldoze stones and soil in front of it, scraping surficial deposits down to bedrock. Melting on the glacier surface creates

Figure 1.1
This glacier was larger during the Little Ice Age, and is melting back. The stream fans out towards the lower right of the picture, just below the dark-colored pile of glacier-transported rocks, called a moraine, marking the maximum size of the glacier during the Little Ice Age. (Photo by Philip Conkling)

streams that drain through great holes in the ice to feed steep, turbulent rivers beneath the ice that further clean the bedrock. A cautious person standing by such a river where it emerges from under the glacier can hear boulders knocking together in the current, and an incautious wader risks a crushed ankle, or worse. The glacier-fed rivers often pile gravel benches or terraces along their paths down the valley. Beneath a glacier, the moving ice picks up boulders and smaller rocks and drags them over the bedrock beneath, scratching and polishing the rock in many places. In other places, the glacier breaks chunks loose and shoves them into the bedrock to make beautiful crescent-shaped cracks. The glacier usually outlines itself with a pile of mixed-up big and little rocks, and the sides of some of those rocks have been worn flat and scratched or polished against bedrock before emerging from beneath the ice. Rocks left behind by a glacier may have been carried long distances in the ice, and many different rock types often are mixed together in deposits of the glacier.

A wealth of other features—layers of mixed-up rock and soil, ponds left by the slow melting of ice blocks buried in the glacially transported material, sinuous ridges marking the former courses of subglacial streams, valley bottoms eroded by the moving ice into broad "U" shapes with waterfalls cascading down as in Yosemite or the fjords of Norway—all these features and more testify to the former presence of glaciers.

But to people who had never seen or studied a glacier, and who may have been raised to believe in a young Earth that had changed little over time, the idea of a vast sea of ice overrunning much of the modern world must have seemed too fantastic for serious consideration. Scotland and England, Scandinavia, the lowlands far beyond the modern glaciers of the Alps, and other parts of Europe bore features from glaciers, but these features were misinterpreted for centuries by people with too little experience in the mountains and too little trust in their own science. The odd boulders carried far from their sources came to be called "drift," based on the mistaken idea that they drifted into place in icebergs unleashed by Noah's flood, perhaps scratching the bedrock on their tumbling paths. For scientists who had not yet visited the polar regions to observe that drifting icebergs produce very different features remarkably unlike those observed across the European landscape, this story sufficed for a time.

We now know that the rounded hills of Scotland, the Matterhorn, the waterfalls of Yosemite, the Great Lakes of the United States and Canada, the Ten

Figure 1.2
Except in especially cold places where ice is frozen to the rocks beneath, glaciers usually erode the landscape more rapidly than rivers or wind. Valleys that are "U-shaped" in cross-section are characteristic of erosion by glaciers, such as the classic Qinguadalen valley that drains into the head of Taserssuaq Lake and was full of ice during the ice age. Note the delta and melt water plume built by sediment carried in the stream fed by numerous alpine cirque glaciers. (Photo by Gary Comer)

Thousand Lakes of Minnesota, and many other features of our planet owe their existence to glaciers. The Chesapeake Bay is a river valley drowned when the melting ice sheets raised sea level. Corals that grow only in shallow water are sitting where they grew, now dead, 100 meters (more than 300 feet) below sea level on the steep sides of many islands, drowned by the rapidly rising sea level as the ice sheets melted at the end of the most recent ice age. The land around Hudson Bay and in Scandinavia is rising, slowly bouncing back after being pushed down by the great weight of the ice of that ice age. The modern world makes sense only if we include the effects of the ice ages.

A few early scientists working near the Alps, including the Swiss mountain climber de Saussure and the German naturalist Schimper, had recognized the evidence that alpine ice had expanded down the valleys in the past. Schimper reportedly conceived of ice ages, and he discussed his insights with the Swiss paleontologist Louis Agassiz. Agassiz, beginning with the 1838 publication of his address at the opening of the Helvetic Natural History Society, is generally credited with formally publishing the scientific proposal for the occurrence of great ice ages in the past, based on extensive studies on and near glaciers, including a significant stint living in a hut on the Aar Glacier.

Once he himself was convinced, Agassiz proceeded to study various glaciated regions with other scientists, showing them the evidence, and often but not always convincing them of the reality of the ice age. (In his later years, Agassiz began discovering "evidence" of glaciation in places where it had not occurred, but that's another story.) Overall, the clear evidence of glaciation came to convince the geologic community of the reality of ice advance. But this discovery raised many more questions that we are still investigating.

Convinced that the ice had once been more extensive, geologists quickly established that many ice ages had occurred, not just one. In places, the ice removed its earlier tracks, as the spreading glaciers of a new ice age bulldozed away the records of any previous ice. But, where a new advance hadn't gone quite as far as an older one, the older deposits could be seen peeking out from beneath the younger ones. Much of the early work was done in Europe, but the evidence is clear in many regions. In the United States, for example, one old ice age reached to Kansas, and a younger one stopped a bit to the north in Nebraska, with still-younger ones extending south to Illinois and then across Wisconsin. Those geologists knew that their history was likely to have gaps.

Figure 1.3
Boulders and other materials carried far from their source by glaciers were originally called "drift", in the mistaken belief that they had drifted into place in icebergs in Noah's flood. Such boulders are now called "erratics." The damage that accumulates in such boulders from exposure to cosmic rays is used to learn when the ice left the boulder, although geologists usually look for even bigger boulders than this one. (Photo by Beverly Walsh)

Four ice ages could have written the record, or forty. The sediments from a small ice sheet would be erased by a larger advance that followed, leaving us puzzled as to what really happened.

Counting Ice Ages

It took a few visionaries, including Wally Broecker in 1970, to solve this puzzle. In the ocean, erosion is rare, with sediments piling up continuously almost everywhere on the sea floor. If a record of ice ages could be found in the ocean, then we could learn what really happened.

Fortunately, there is a record. The story starts far from glaciers, with the realization that there is a little bit of "heavy water" naturally mixed into the ocean from the presence of heavy isotopes of oxygen (oxygen-18 instead of oxygen-16) or hydrogen (deuterium instead of "normal" hydrogen). Roughly one of each 500 water molecules weighs a bit more than its neighbors because it has an extra neutron or two in one or more of its atoms. These heavy molecules are still water, but they're heavier. It is not surprising that these heavy molecules don't evaporate quite as easily as the "ordinary" light ones. In warm places and times, water that evaporates from the ocean returns as rain or river flow rather quickly. But during an ice age, a whole lot of water that evaporated from the ocean falls as snow on land to accumulate into vast quantities of ice. Sea level is lowered roughly 100 meters (more than 300 feet). Because the heavy water is preferentially left behind in the ocean, heavy molecules are a bit more common in ocean environments during an ice age, while the end of an ice age causes the heavy water to be diluted by the return of water from the melting ice.

Many creatures live in the ocean and build shells. The chemicals in these shells include oxygen, where the ratio of heavy to light oxygen in the shells reflects the ratio of heavy to light oxygen in the ocean water in which the shells grew. Paleo-oceanographers—the scientists who study the history of the oceans—have found it especially useful to study the shells of small creatures called foraminifera, in particular focusing on those "forams" that make shells of calcium carbonate and that live far enough out to sea that waves and landslides don't mix up the mud that accumulates on the bottom.

The procedure for studying forams is fairly straightforward. Scientists go out in a ship with a drill on it, shove the drill—a high-tech glorified pipe—into the sea bed, pull up a core from the mud, open it up, and sort out the foram shells layer by layer. The ones on top are the youngest, while the deeper ones are older. (Ages can be estimated in various ways, with the estimates now quite accurate.) If you measure the ratio of heavy to light oxygen in the shells of forams, you are measuring how much light water had been removed from the ocean to make ice sheets at the time the shell grew. The core becomes a sort of tape recorder that tracks the history of ice ages. There is an extra complication, because the heavier oxygen goes into shells more easily when the water is colder. But, the climate is colder during ice ages, so both growth of ice sheets and cooling of the climate favor heavy oxygen in shells.

Sediment cores from all the world's oceans have now been studied many times, and they tell a remarkable and remarkably consistent story. Over the last approximately one million years, ice grew slowly (although with a few "bumps" of temporary shrinkage) for about 90,000 years, then melted over about 10,000 years (again with some bumps), and then repeated this 100,000-year cycle. Bumps are spaced about 41,000 years apart, and also about 19,000 to 23,000 years apart. Earlier than the last million years, there were a few million years in which the ice did most of its growing and shrinking on the 41,000-year schedule, with the faster and slower changes notably smaller. If we go too far back in time, we find the dinosaurs lived on a very different, alien-to-us world too hot for ice, and perhaps too hot for us in a lot of places.

Cycling with Milankovitch

Remarkably, the timing of the ice ages was predicted decades before the data were available from sea-floor cores. The predictions came from calculations of Earth's orbit.

Among the great triumphs of physics is the ability to understand and predict the motion of things, including the planets. The predictions have been very successful; rocket scientists really can hit the moon of another planet, or a comet, or a tiny orbiting dot of a space station, with amazing accuracy. This isn't that easy, either, because Earth, other planets, moons, comets, and satellites are doing so many complicated things at once.

Think about a child's top, spinning on the table. The red ball perched atop the top doesn't stay perfectly in one place; it wobbles around in a circle. This motion is called *precession*. Earth does this, too, with the North Pole tracing a small circle over about 26,000 years. The interaction of that 26,000-year circle with all the other orbital dynamics that are going on gives rise to the ice-age variations between 19,000 and 23,000 years.

The child's top slowly tilts more and more as it circles, and eventually the top falls over. Fortunately, Earth doesn't fall over. Instead, Earth's tilt increases and then decreases, just a little—about 3 degrees total. One full tilt cycle takes 41,000 years. The tilt is formally called the *obliquity*. The precession and obliquity of a child's top are linked to gravity pulling down on the toy. Earth's rotation causes it to bulge a little around the equator, and the sun, moon, and other planets pulling on that bulge give rise to Earth's precession and obliquity.

The orbit of Earth about the sun is not a perfect circle but an out-of-round ellipse. The sun is not in the center of that ellipse, but instead is shifted slightly along the longest axis of the ellipse. The shape of that ellipse changes, going from more nearly circular to more squashed and back over about 100,000 years. The shape of the orbit changes primarily because of the gravity of Jupiter, which gives a little gravitational tug every time Earth laps that great planet with our faster orbit in the race around the sun. The out-of-roundness of the orbit is called its *eccentricity*.

Several brilliant astronomer/mathematicians figured out how these features of Earth's orbit work and how they affect sunshine and our seasons. Adhemar in 1842, just a few years after Agassiz established the theory of ice ages, suggested a role for precession. Croll in 1875 added eccentricity and obliquity. Then the Serbian mathematician Milutin Milankovitch greatly expanded this work between about 1920 and 1941, calculating the variations over time in solar radiation at the top of the atmosphere at different latitudes—a remarkable feat before electronic computers. As a result, Milankovitch is usually given most of the credit, and so the various cycles are often called Milankovitch cycles.

If the North Pole stuck straight up, the sun would barely graze it; if you think of the North Pole as a bald spot on the head of the planet, that bald spot would never get a sunburn. But the obliquity (tilt) of Earth's orbit allows the sun to shine on the pole more directly during summer, giving the bald spot a nice tan. As the obliquity increases, the equator is not pointed as directly toward the sun and so the equator gets less sunshine, while the poles get more.

Earth's seasons arise from the obliquity in combination with its the orbit about the sun. Because the Milankovitch cycles take tens of thousands of years to make a real difference, these cycles have essentially no effect over the short time of a single year. You can think of the tilt of the North Pole as not changing at all over a few years or even a few centuries. In December, Earth is on one side of the sun, with the North Pole tipped away from the sun while the South Pole is being warmed by the sun's rays. By June, Earth has moved halfway around the sun, but Earth's tilt is still the same, so in June the North Pole is sun-tanning while the South Pole gets no sun.

Because Earth's orbit is slightly eccentric rather than round, our distance from the sun varies slightly from year to year. For the last couple of millennia, and for the next couple of millennia, the north has summer and the south has winter when Earth is far from the sun. Thus, for people in the northern hemisphere, summer and winter are not quite as different as they might be. However, about 11,000 years ago—half of a precession cycle—northern summer and southern winter happened when Earth was closest to the sun on our eccentric orbit, so people in the northern hemisphere had especially warm summers and colder winters. Over 11,000 years the eccentricity has not changed a whole lot—the eccentricity cycle takes 100,000 years. Both 11,000 years ago and now, the eccentricity of the orbit has not been especially high. At times further in the past, and at times to come well into the future, larger eccentricity led to, and will lead to, larger changes in the distance to the sun, and thus larger changes in the seasonal distribution of sunshine in response to precession. In addition, the obliquity/tilt has been decreasing recently, shifting a little sunshine from the poles to the equator.

These orbital features have only a tiny effect on the total sunshine reaching the whole Earth. But the change in sunshine at a particular latitude and season

can be more than 10 percent, making a big difference in annual temperature that lasts for thousands of years.

Milankovitch Summarized

Thus the features of Earth's orbit described above move sunshine around, from north to south, poles to equator, summer to winter, and back, over tens of thousands of years. Milankovitch calculated these slight changes, noted that the total change in sunshine at a place during a season can be more than 10 percent, and suggested that this should cause ice ages, with spacings corresponding to the main orbital features of about 19,000 to 23,000 years, 41,000 years, and 100,000 years.

Decades later, when the drill ships and the foram-analyzers and the other tools of the paleo-oceanographers yielded good data on the history of ice ages, exactly these cycles appeared in the ice-age story. When such a prediction is confirmed so beautifully from an independent source, we can have high confidence that the ideas behind the prediction are correct. Thus we can say confidently that slight changes in orbits cause ice ages.

However, Milankovitch thought that the ice ages would switch from north to south, running away from the intense summer sun in response to the 19,000 to 23,000 year cycle. Instead, the whole world cooled at the same time and then the whole world warmed together. And the world seems to have primarily followed the amount of sunshine in the northern summer. Milankovitch had many things right, but a lot of fine-tuning has been needed to update his ideas. For example, Earth seems to have switched from 41,000-year-spacing to 100,000-year-spacing of ice ages because the ice sheet on Canada was able to get sufficiently bigger after it swept all the soil off large parts of Canada that the ice could survive the shorter-lived warmings caused by the orbits; it is hard to build a big ice sheet atop slippery soil, and easier on hard rock after the soil is gone.

Figure 1.4
Many of the coastal highlands of Greenland are crowned by ice caps, as shown in this aerial view of West Greenland. During the ice age, the ice flowing from these highland ice caps deepened the spectacular glacial valley, which has been flooded in its lower reaches to form a fjord. (Photo by Philip Conkling)

Changes in Carbon Dioxide

When changes in the orbit reduce sunshine in the north, snow can survive the summer on highlands around Hudson Bay and in parts of northern Eurasia, while Greenland is almost always capped in ice. When changes in the orbit reduce sunshine in the south, there is already ice on Antarctica, and the next land on which ice could grow is so far toward the equator that only the highest peaks may stay cold enough in summer to grow glaciers. Thus, it is not surprising that the north proves to be more influential than the south in controlling the climate.

Our understanding of the climate initially agrees with Milankovitch insofar as an ice sheet in the north should make the north colder but not have much of an effect on the south. If you cool the north long enough to grow ice sheets, other changes occur that make the cooling somewhat bigger, which climate scientists call a *positive feedback*. Snow and ice are highly reflective, so a bigger ice sheet reflects more of the sunshine back to space, making the climate colder. The ice grinds up rocks to make dust, and when the dust blows around, it blocks the sun a bit. Grasslands and tundra replace darker forest as the ice grows, increasing reflection of sunlight more, especially because tundra can be easily buried by snow whereas dark trees stick above it. So if changing orbits cause a decrease in the sunshine in the north and allow summer snow to survive, the climate can really cool in the north. But the dust mostly stays in the north, the largest vegetation changes occur the north, and the reflection of sunlight occurs disproportionately in the north. Both our physical understanding and our climate models indicate that little of the cooling influence will cross the equator; decreasing northern sun while increasing southern sun should cool the north but warm the south.

Yet, the climate in the south has changed in the same way as the climate in the north. Why? Ice-core records (which we describe more fully in the next chapter) show clearly that greenhouse gases also changed with the northern sun. Soon after the start of warming we observe a rise in greenhouse gases. Similarly, as cooling kicked in and the ice grew, carbon dioxide and methane and nitrous oxide responded, dropping soon after the temperature began falling.

Physical understanding tells us that an increase in greenhouse gases must have a warming tendency. We can test this against history, and it works very well. If the physical understanding is correct, then we can successfully explain

Figure 1.5
A great glacier sweeps towards the coast in east Greenland. If this glacier were to melt, it would reveal a glacial valley much like those in figures 1-2 and 1-4. (Photo by Philip Conkling)

the history without cheating—the south marched through the ice-age cycles in step with the north, because the greenhouse gases affect both poles. If the physical understanding is wrong, then no one has any idea how the ice ages worked—no successful explanation of the size of the temperature changes over ice-age cycles has ever been presented that ignores the greenhouse-gas changes.

A complete explanation of the changes in greenhouse gases is a bit beyond the scope of this chapter, and indeed some questions remain that science has not fully answered. Methane and much of the atmosphere's nitrous oxide come from marshy areas on land. Since most of Earth's land is in the northern hemisphere, the cooling of the ice ages dried many of those wetlands and so reduced the sources of those greenhouse gases.

Some carbon dioxide shifted from the air into the ocean during ice ages, and the carbon dioxide came back out into the atmosphere during the warm interglacials between ice ages. These changes in carbon dioxide were slower and smaller than what we humans are causing now, but were still substantial. Many different processes were involved. For example, plants that live in the surface ocean are primarily made of carbon dioxide and water, which are combined using the sun's energy for photosynthesis. When the plants die, some of their remains sink (the plants may be eaten and then packaged as fecal pellets before they sink). Animals living in and under the water use this sinking material for food, liberating the carbon dioxide in it.

Meanwhile, back at the surface, the spray and bubbles of breaking waves mix the air and water well, so the carbon dioxide that sinks into the deep ocean after being taken from the surface ocean by plants is replaced by more carbon dioxide dissolving in the ocean from the air. Eventually, the circulation of the ocean brings the deep waters back to the surface, especially in the Southern Ocean around Antarctica, where the deep water releases its excess carbon dioxide to the air, closing the cycle.

Anything that increases plant growth in the surface waters, or that decreases the ability of the deep waters to get to the surface and release their store of carbon dioxide, will lead to lowering of atmospheric carbon dioxide over centuries. During the ice age, more wind-blown dust meant more fertilizer reached the ocean to grow more plants. The lower ice-age sea level allowed rivers to dump their mud closer to the open ocean rather than being stuck at

Figure 1.6
The muddy streams emerging from beneath Sermeq Kangigdleq glacier, southern Greenland near Julianehab, attest to the erosive ability of the ice. The lakes in front of the glacier formed as buried ice melted, after being left behind during rapid retreat. Ice probably exists in places near these lakes beneath the mud. (Photo by Gary Comer)

the head of Chesapeake Bay and other such bays, again tending to fertilize the ocean more during ice ages. Expanded sea ice during ice ages tended to block the escape of carbon dioxide from the deep waters. By these and other mechanisms, the growth of ice sheets in the far north lowered atmospheric carbon dioxide, which then helped make the ice ages go global.

A Few More Puzzles

Telling the story of the great sweep of ice ages over tens of thousands of years took an immense amount of scientific effort over centuries, and active research still continues. The ice sheets melted rapidly with warming, and they sometimes belched out great surges of icebergs, worrisome when we consider that we are warming the world with ice sheets still perched on Greenland and Antarctica and capable of raising sea level if they melt rapidly. The climate changes of the ice ages provide a wonderful laboratory for testing our understanding of the climate system, and the success of our climate models in re-creating what happened increases our confidence that we understand most of the key features of how Earth's climate works, and thus that we can trust our projections of future climates.

But in learning the history of ice-age climates, even more peculiar things kept showing up. The warming from the last ice age wasn't smooth; it was interrupted by cold events, and especially by a prominent millennium-long event called the Younger Dryas, described in detail in chapter 4. A small tundra flower with the scientific name of *Dryas* had migrated south just ahead of the ice sheets, and its remains showed up in bog sediments from just beyond the ice; the remains of *Dryas* disappeared from bog sediments deposited when the warming from the ice age began, but then reappeared, evidence of another cooling. The start and end of the Younger Dryas were fast, fast enough to matter a lot in a single human lifetime, and maybe over the single term of a politician. The orbits changed far too slowly to explain such a feature as a Younger Dryas event. So what was it, why was it so abrupt, and could such a thing happen again?

The warming from the ice age had been enough to melt the ice on Eurasia and Canada. Actually, there are little ice caps still perched on highlands of some of the islands of Arctic Canada, and a few of those little ice caps contain

Figure 1.7
Icebergs and broken-up sea ice offshore of the fishing village and tourist destination of Illulisat, Greenland. (Photo by Gary Comer)

a thin layer of ice left from the great ice-age ice sheet. So, the ice sheet didn't really disappear completely; it just shrank a whole lot. But Greenland, with ocean on all sides to supply moisture to snowstorms, managed to hang on to most of its ice as the ice age ended.

Part of the reason that Wallace Broecker, George Denton, and many of our other colleagues have spent so much time studying climates of the past directly relates to the goal of trying to understand what will happen in the future as our atmosphere continues to absorb heat-trapping gases such as carbon dioxide, methane, and nitrous oxides. Earth's climates of the past enable us to test our climate models. That is, if we understand the fundamental processes that control how Earth's orbit affects our atmosphere, oceans, and vegetation systems and how all of these parts of a very complicated biosphere interact with each other, we can start with known conditions of the past and run the climate models forward to today. If our climate models produce the ebb and flow of ice ages when we alter fundamental processes, we have more confidence that we understand which processes control climate change. We also have more confidence in the result of running the models forward to ask what happens as conditions in the atmosphere, ocean, and landscape change

How much more warming could Greenland's ice sheet take before following all the rest of Earth's northern ice sheets into watery oblivion? Will we cause that much warming in the near future? These are rather large questions, and they will occupy much of the rest of this book, just as they are occupying much of the research careers of the authors.

Figure 1.8
The high-altitude plateaus and peaks of the coastal mountains are cold and snowy enough to keep their ice caps, such as this one on Milne Land. An ice core through the Renland ice cap, near where this photo was taken, has contributed to understanding of the history of Greenland. (Photo by Gary Comer)

ROSETTA STONES FROM THE GREENLAND ICE SHEET

The Discovery of Abrupt Climate Change

A master detective notices the one clean square surrounded by thick dust on the shelf, and knows that the box with the evidence has been taken away. An archeologist uses brush and trowel, and layer-by-layer unearths the history of the long-gone people. And a polar explorer rises in the morning to see that snow from last night's storm has buried the tracks from yesterday. The polar explorer is a friend of the archeologist, has read the tales of the great detective, and knows instinctively that the layers of ice beneath those buried tracks must be a history book, too.

Alfred Wegener, who is famous for his exposition of continental drift, was a polar explorer and meteorologist who lost his life exploring Greenland in the early part of the twentieth century. During Wegener's 1930–1931 expedition, Ernst Sorge wintered-over in a snow cave at Eismitte (literally "mid-ice" in German) near the top of the Greenland ice sheet when temperatures plummeted to -65°C (–85°F). Sorge hand-dug a 15-meter-deep (50-foot) pit, mapped the annual layers of snow, and, without killing himself by falling into the hole, proposed what we now know as Sorge's Law, describing how ice density changes with time.

By the 1950s, Greenland began to be probed more widely and more deeply by scientists especially associated with what we now know as the Cold Regions Research and Engineering Laboratory (CRREL) of the US Army Corps of Engineers. Under the direction of the Swiss-American Henri Bader, pioneers such as Carl Benson, Chester Langway, and Tony Gow first succeeded in drilling through the ice sheets of Greenland and Antarctica and began reading the climate record. International research collaboration also grew rapidly; collaborations in the 1980s between Willi Dansgaard of Denmark and Hans Oeschger of Switzerland proved especially important.

These were truly heroic times in exploration. Carl Benson, for example, spent four years criss-crossing Greenland in a tracked vehicle (a "Weasel"), hand-digging snow pits to map the accumulation on the ice sheet. The first complete penetration of the Greenland ice sheet was finished in 1966 after six years of effort, and several drills, at Camp Century in northeast Greenland. The joint US-Danish-Swiss Greenland Ice Sheet Project (GISP), with Langway, Dansgaard, and Oeschger leading the scientific inquiries, followed this initial effort. Seven years of preliminary site-selection and analysis work led to coring the ice sheet down to bedrock on the southern dome of the ice sheet, at Dye 3, between 1979 and 1981. Further projects have followed; most notable for us are the twin cores from the summit of the main dome of the ice sheet, the primarily European Greenland Icecore Project (GRIP), and the Greenland Ice Sheet Project II (GISP2) under the direction of chief scientist Paul Mayewski. Beginning in 1989, Richard Alley spent a lot of time at GISP2 and rode the success of that project to academic tenure.

Ice coring is quite simple in principle. If you take a pipe with teeth on the end and spin it against the ice, you cut a core out of the ice. After you've gone far enough, if you break the core, grab it in the pipe, and bring it to the surface, you have a sample to study. Run your drill back down the hole, and repeat. In Greenland, once you have about 3 kilometers (almost 2 miles) of cores, you will hit bedrock, maybe drill into that a bit for good measure, and go home with your samples.

Of course, it isn't quite that simple. Your pipe will get really long and heavy unless you hang it on a cable, so you'll need a cable and a winch, a drill tower, and some springs to press against the side of your hole so that the drill doesn't stop while the cable spins. Anyone who has ever drilled or sanded knows that

Figure 2.1
Greenland ice cap flowing toward head of the Daugaard–Jensen glacier, which drains into the head of Scoresby Sound. (Photo by Philip Conkling)

sawdust gets in the way and must be removed. For ice coring, the "chips" are usually moved up a spiral ramp and dumped in a space above the core, to be removed from the hole, although one drill dissolved the chips in antifreeze, while other drills work by melting rather than cutting. An empty hole a few hundred meters deep will squeeze closed rapidly under the weight of the surrounding ice, so the hole must have a fluid in it to keep the walls from caving in, and the upper part of the hole is generally cased with something to keep the fluid from draining away into the spaces in the snow that has not yet been squeezed to ice.

Ice-core drilling has been highly successful. But those who have sat at the surface waiting for the next core know that real genius, hard work, and maybe a bit of luck are required in order to succeed.

A BIG Pile of Ice

A little background on ice sheets makes the interpretation of ice cores easier. A glacier or ice sheet forms where snowfall exceeds melting over the years, allowing the snow to pile up. In warm places, the snow may become water-saturated during the summer and refreeze in the winter, making solid ice in a single year. But the water may move dust around, trap gases at ratios that don't match those in the atmosphere, and allow bacteria and other microorganisms to live in abundance, creating waste products that change the chemistry of the ice. In short, the climate records are altered in ice that has experienced a lot of melting, and hence those glaciers that melt and refreeze often are usually not the targets for ice-core studies of past climate.

Instead, colder places are the main target. In the central regions of Greenland and Antarctica, melting does not occur even on the warmest summer day. (Well, over the last few millennia, central Greenland has generated a little bit of refrozen melt water in scattered years averaging about 300 years apart, but no such refrozen layers are known from central Antarctica.) In such really cold places, snow piles up on top of old snow. The building weight squeezes the snow beneath, slowly pushing the air out and moving the ice grains closer together. Eventually, the air spaces are pinched off, trapping bubbles. The old snow

Figure 2.2
An ice sheet forms where snowfall exceeds melting over the years. Pictured here is the Greenland ice cap looking south from Kangerlussuaq or Sondrestrom. (Photo by Gary Comer)

is called *firn*, and the firn becomes ice when the bubbles form. The formation of ice from snow and firn may take centuries or even millennia, especially in very cold places where the snow piles up really slowly.

The snow on Greenland or Antarctica does not pile up higher and higher until the world becomes top-heavy and rolls over. Instead, ice sheet flow moves ice to warmer places to melt, or to the ocean's edge to make icebergs that drift away and melt. Usually, an ice sheet is nearly "in balance," with annual snowfall creating new ice about as rapidly as melting and iceberg calving remove old ice.

All piles tend to spread under their own weight. Try to make a pile of water in the middle of a table, and you very quickly will observe this spreading tendency, the pile becoming very wide and not very thick. Use warm maple syrup and the pile will spread more slowly. With cold maple syrup, the spreading is even slower. An ice cube sitting on your table is likely to melt long before you can observe any spreading, making water that spreads much more easily. Really strong piles may not spread at all; the strength of wood holds a table together so that it does not spread out very wide and thin.

But if you have a 3-kilometer-thick pile of ice covering a continent, or the world's largest island—Greenland—the pile will spread. This is actually fairly easy to measure. Put a stake in the surface somewhere up on the ice sheet, and use a GPS receiver to figure out where the pole is. When you come back with your GPS receiver a year or two later, you will find that the stake has moved.

It might seem odd that while you are walking around on the ice, driving a snowmobile on it or landing a plane on it, the ice underneath is actually flowing. Solids typically become soft enough to flow slowly when heated close to their melting point, even while still solid. A blacksmith can form a horseshoe from a bar of iron with heat and a hammer and yet not melt the iron. A big enough pile of iron, heated just below its melting point, would spread slowly under its own weight. Ice seems cold to a polar explorer, but it usually is as close to melting as is the iron worked by a blacksmith. So, the ice can flow.

On the flowing pile of ice, your GPS receiver will show you that stakes on the east side of the ice sheet are moving east, toward the nearest coast. Stakes on the west side of the ice sheet will be moving west, while stakes on the north side head north, flowing away from south-side stakes that move south. The motion is always downhill, away from the direction that the upper surface slopes, so the central point from which all of the ice moves is also the highest point.

Figure 2.3
The dark blue ice streaks in this iceberg are formed from refrozen melt water. (Photo by Gary Comer)

Figure 2.4
Calving tidewater glacier at the head of Kangersuneq qingordleq glacier in Northwest
Fjord, Scoresby Sound; the absence of icebergs indicates the flow of this glacier is quite
slow. (Photo by Gary Comer)

Figure 2.5
Flow of the ice sheet towards the ocean opens crevasses in many places. Flow past rocky regions may pick up enough rocks to form a medial moraine, such as the curving brownish stripe in the center-right of the photo, and the stripes on the glacier in figure 2.4. (Photo by Philip Conkling)

(Flow can be a bit more complicated than this; in Greenland, ice spreads away from the main dome and from a smaller southern dome, turning to head toward the coast where these flows run together.)

If east-side ice moves east, west-side ice moves west, and there is no giant crevasse in the middle, then a single layer of snow or ice must be getting wider and thinner, just as a pile of water or maple syrup spreads and thins on a table top. The layer of snow from 2,000 years ago will have been spreading and thinning for two millennia. The layer of snow from a more-recent year (say, 1,000 years ago) will have had less time to spread, and so will not have thinned as much. For the central region of an ice sheet that has come into balance, the thinning will cause the uppermost layer to move downward just enough in a year to make room for the next year's snowfall, so the ice sheet neither grows nor shrinks.

Because deeper layers have been thinned more, the age of the ice increases very rapidly as you drill toward the bed of an ice sheet. Crudely, the thickness of a layer is reduced by half as you drill halfway from wherever you are toward the bed. This cannot be exactly correct (since it implies that the ice is infinitely old at the bottom, for example, so a little more care is required to account for complexities there), but this works well enough for the discussion here.

Figure 2.6
The narrow ice ridges between closely spaced crevasses are called "seracs," such as these spectacular examples on a glacier flowing off the ice cap into Melville Bay on the northwest coast of Greenland. (Photo by Gary Comer)

Reading the Record

A three-kilometer-long ice core, even in meter-long pieces, is an ungainly history book. And the history is written in unfamiliar script. The first step in reading that unfamiliar script is to learn the age of the core—something like checking out the page numbers. The sun shines in the polar summer but not in the winter, so summer snow is "cooked" by the sunshine even without melting, creating larger crystals and causing other differences. If you place the core on a glass table with a light shining through, you can see the layers as faint "tree rings" or "ice-rings" of history.

Many other indicators of summer and winter also exist. The hydrogen peroxide produced in the air by the summer sun falls out rapidly into the snow,

while none is made in the winter. Winds from beyond the ice sheet bring more dust and sea salt in some seasons than others. Chemical analysis can thus pick out individual years. The chemicals affect the electrical properties of the ice, so rapid identification of years is possible with clever electrical techniques.

But how do we know that the "years" are really years, and not random layers, or individual storms, or sunspot cycles, or something else? The first step is to watch layers form, understand the processes, and rely only on properties known to change annually. Using many different indicators helps. These are best if they involve "blind" intercomparisons—you count, then I count without peeking to see what you got, and then we compare. If we agree, great; if not, back to the drawing board.

More important, as far back as written history, we know the ages of big volcanic eruptions because people have tended to write the story after a blast. Eruptions that reach into the stratosphere spread ash worldwide, and little bits of that ash are trapped in the ice cores. Chemically "fingerprinting" the ash reveals the volcanic source. A prominent marker in Greenland is the devastating 1783 eruption of Laki in southern Iceland, but many other such eruptions can be picked out with confidence. Comparing the known age to the number of years counted using the different techniques shows how accurate the counting is. Experience demonstrates that in "good" cores with enough measurements, and where there has been plenty of snow accumulation—with an annual layer thicker than the height of a snowdrift—it is not too hard to have 1 percent accuracy. If we tell you that this section of core contains 100 years, it almost always contains 99, or 100, or 101, but not 112 or 83 or some similarly large error. Even zero-error dating is sometimes possible.

With the age of the layers in hand, we can estimate the snowfall rate from the thickness of an annual layer. A correction is needed for the thinning from ice flow, and for the amount of air trapped in the ice. The trapped-air correction is easy. The correction for ice flow ranges from easy in the youngest ice to nearly impossible in the oldest ice. Everything we'll discuss here relies on fairly easy corrections.

Ice with a higher concentration of dust, or pollen, or sea salt, or micrometeorites, or anything else that blew through the air, may be telling one of two stories: either the air was delivering more of that material, or the air was delivering less snow and ice to dilute the material. Once we can calculate snowfall from the thickness of annual layers, however, the dirtiness of the ice reveals the dirtiness of the air. We thus can learn about the delivery rate of dust, sea salt, micrometeorites, and whatever else is in the ice.

Most of Earth's dust is produced from certain places—the Sahara and Gobi deserts make lots more dust than does the Amazonian rain forest. And the dust from the different big sources is chemically distinct, so that an expert can tell whether a given dust sample is from the Sahara, or the Gobi, or is a mixture of the two. In this way, an ice core can reveal how much dust was coming from where, and we can make inferences about atmospheric winds and desert conditions.

When snow is squeezed to ice, bubbles of air are trapped. Those bubbles are held in the ice, and they provide the only reliable archive of old air—fossil air, you might say. We test the reliability in many ways. Samples formed very recently have almost exactly the same composition as the atmosphere above the ice at that time. Samples taken from different ice cores collected from places with different snowfall rates, temperatures, and impurities tell the same story, as long as the ice is not too dirty or warm. However, in really dirty, warm ice—say, ice cores from small high-mountain tropical ice caps containing dead flies that blew in from the surrounding land—the trapped gases are shifted from the atmospheric value in the way we would expect if dead flies were decomposing in fairly warm ice. We thus understand how the measurements work, where and when they work, and what "breaks" them so they don't work, and we can use the data produced from good samples with confidence.

Greenland's Ice Core Record

While lots of Antarctic ice cores provide beautifully consistent histories of atmospheric carbon dioxide, Greenland cores give "noisy" records. Although the dead-fly problem is not found in Greenland, there is just enough volcanic acid and calcium-carbonate dust in the ice cores of Greenland that some samples end up with a little bit of extra carbon dioxide. Geologists know that calcium carbonate is identified in the field by the carbon dioxide that bubbles off when a little acid is added, and the ice of Greenland is just dirty enough to produce a little of this.

Dust, pollen, and flies don't stay in the air very long. Most dust falls out of the air in a week or two, before the wind can carry the dust around the planet. Thus, ice cores in Antarctica and Greenland tell different stories, because they are looking at dust from different places. However, most gases stay in the atmosphere much longer than the year or two needed for winds to mix the gases around the planet, so ice cores in Greenland and Antarctica trap air with the same composition. Methane is an interesting in-between case. Methane typically lasts only about a decade in the atmosphere before being converted to carbon dioxide and water. Most natural methane is formed in the wetlands of tropical and northern lands, so the concentration measured in Greenland nearer these sources of methane is expected to be just a little higher than in Antarctica, and this in fact occurs.

Extracting the temperature record in the past is one of the most interesting parts of ice-core science. An ice core reveals temperature on the ice sheet, not the global temperature, so other techniques must be used in places without ice. But, an ice core has lots of different ways to reveal local temperature, so the temperatures estimated from ice cores are especially reliable.

Some "paleothermometers" are fairly easy to understand. For example, if a little bit of melt water is produced on a rare hot day and then refreezes, a layer is produced with almost no bubbles. These "melt layers" are very easy to see in normal bubbly ice. In central Greenland, changes in how frequently these occur provide information on past summer temperatures.

A frozen roast placed in a hot oven takes a while to warm up inside, and a cautious cook uses a meat thermometer to track the internal temperature until it is high enough. On the other hand, if you tell the cook how long a roast has been in the oven, how big the roast is, and how hot the oven is, the cook may be able to tell from the temperature in the middle whether the roast was frozen or just chilled before being put in the oven, because the frozen roast takes longer to warm.

The same principle applies to ice sheets—the ice 1–2 kilometers down in central Greenland is colder than the ice near the surface because that deeper ice has not yet finished warming from the cold of the last ice age. With knowledge of the thickness of the ice, the flow of the ice, the snowfall rate, and the heat coming up from the rock beneath, the temperature of the ice 1–2 kilometers deep tells us how cold the ice age was.

Reading history with a glorified meat thermometer stuck into the ice sheet provides highly reliable values, but the record of short-lived temperature events that happened a long time ago is smoothed away. Fortunately, several other techniques can detect old short-lived events. These techniques are harder to explain, but not harder to apply.

As described in chapter 1, if you grabbed a handful of ocean water and looked at the water molecules one by one, you would find that about 1 in 500 is slightly heavy—there is an extra neutron or two in one or more of the atoms in the molecule. Just as adding a little extra weight wouldn't really change who you are, a water molecule with an extra neutron or two is still water. But, just as a heavier person might have a more difficult time jumping up, and be more tempted to sit down, a heavy water molecule is a bit less likely than a light one to evaporate from the ocean, and the heavy ones that do evaporate are a bit more likely than the light ones to condense and fall back down.

When a cloud moves in from the coast over Greenland, the heavy water molecules more readily condense and fall as snow near the coast. As the cloud slowly runs out of heavy ones, the snowfall comes closer and closer to being all light ones. Importantly, cooling the air over the ice sheet causes more of the water molecules to fall out on the way to the ice-sheet center, taking more of the heavy ones and forcing the next snowfall to be even richer in light ones. So, the ratio of heavy to light water molecules in an ice core is a thermometer. We call this the *isotopic ratio* of the ice.

We have additional paleothermometers, based on the isotopic composition of trapped gases, or on shifts in that isotopic composition after abrupt climate changes, or on the size of bubbles, among other indicators. Some of these indicators provide histories of very short-lived temperature changes, whereas others smooth the history a little bit. Putting all of the temperature indicators together provides a wonderful history of the surface temperature on the ice sheet and provides enough redundancy that we can say with high confidence that the reconstructions are accurate.

Figure 2.7
Clouds move in over Julianehab Ice Cap at the head of Sondre Sermilik Fjord in southern Greenland. The greenish color of the water is caused by suspended silt (glacial flour), which is generated by subglacial abrasion and delivered primarily by subglacial streams. (Photo by Gary Comer)

Into the Depths

Returning to the early 1990s in central Greenland, Richard Alley is perched atop the ice sheet at the very summit, with other scientists and drillers boring relentlessly into the ice sheet as part of the Greenland Ice Sheet Project II (GISP2). Twenty-eight kilometers to the east, a European team is doing the same at the site called Greenland Icecore Project (GRIP).

The sun swings around the sky, higher and lower without ever setting. During a sunny day, the temperature hovers just a few degrees below freezing, while at night the cold settles in as the air cools. With 3 kilometers of ice piled on bedrock that is now near sea level, the air is thin. New arrivals gasp for breath the first few days and sometimes ask for a handy oxygen bottle. The view beyond the camps is of snow; sparkling, untracked, and renewed frequently by the strong storms that blow through, piling drifts behind everything human-made and sifting snow into any unsealed cracks.

At the GISP2 camp, a blue "big house" perched on massive stilts lets the snowdrift beneath the galley, communications center, and general meeting space. Tents and canvas Quonset huts provide berthing for the scientists, drillers, and support staff. Often more than 50 of us are in residence. A caterpillar tractor ferries snow to the melter to supply water, and grooms the runway for the ski-equipped LC-130 Hercules heavy-lift aircraft that the US 109th Air National Guard flies in support of the camps. The whine of a winch raising the drill comes from beneath a giant white geodesic dome that protects the drillers from the worst of the weather, with a 30-meter drill tower sticking out the center of the dome.

Next to the dome, out of sight, is the under-snow laboratory. The drill collects 6-meter-long sections of core, which are dried from the fluid, cut into 2-meter-long sections, and placed in special trays that the scientists roll along a wheeled trackway on one wall of the lab. A band saw slices along the core, removing a thin piece of ice that is bagged for later isotopic analysis. A second, thicker slice is removed, taken into side labs, and split into "sticks" for chemical, dust, and other analyses.

Next, the core is wheeled to the electrical station, where two electrodes about 1 centimeter and 1000 volts apart are dragged along the core. A green

Figure 2.8
The lower-elevation regions experience extensive summertime melting, often forming lakes in low places on the ice. Repeated freezing and thawing of these lakes, and snowfall on the lake ice when it is frozen, can create fascinating patterns, as shown here. (Photo by Gary Comer)

line on a computer screen rises and falls with the seasons, climbing especially high for the acids of a major volcanic eruption, and plunging low in the smoke from a major forest fire somewhere upwind in Canada or the United States. After producing the electrical map, the core is examined by a geologist who notes annual layers, any disturbance to the layers, any fractures or other issues with the quality, and anything else of interest. Next, pieces of core are scanned with a laser beam to provide another record of annual layers. The core is then packed, first in a plastic bag, then in a silvery tube, then in an insulated box, and lowered into a deeper, colder layer of the laboratory for storage.

Later, the cores will be hauled to the surface and loaded on "cold decks" of the Hercules aircraft (leave the heat off!), flown back to the States, and taken in freezer trucks to the US National Ice Core Laboratory in Denver, or to various other labs where additional analyses are made. And from this will come a remarkable record of Earth's climate extending back more than 100,000 years.

Surprises in the Ice Core

A lot of the results from the ice cores are highly reassuring, confirming and extending our understanding of the climate system. But, the ice cores also show some very unsettling things, which lead up to the big surprise.

A big volcanic eruption puts a lot of sulfur dioxide into the stratosphere. This changes to particles of sulfuric acid, which block the sun and create a little cooling for a year or two. The cold and the acid both are recorded in the ice cores. If volcanoes could get organized, they would make a big difference in the annual climate of Earth, but there is no way for a volcano in Alaska to tell a volcano in Indonesia that it is time to erupt. So, the volcanoes just make "noise" in the climate; the time of the next volcanic eruption is unknown, but we do know there will be another one, and that a bit of short-lived cooling will follow.

When the sun is a little more active, not only does it give Earth a little more energy, but it also tends to shield Earth a little more from cosmic rays produced outside the solar system. Cosmic rays break atoms in the air, making odd nuclides such as beryllium-10 that fall on the ice sheet. Ice cores thus can be used to help reconstruct the history of the sun's activity. The records show that the sun has experienced small fluctuations, tracking the 11-year sunspot cycle, and that more slowly the climate has responded. If the sun ever had big, fast changes, we would be unhappy, but the records show the sun to have been a faithful companion with only small fluctuations.

With some help from scientists working on lava flows, the ice cores also show that cosmic rays themselves are not important to the climate, nor is Earth's magnetic field. The cosmic rays that make beryllium-10 are steered away from Earth by the solar wind, but also by Earth's magnetic field. The magnetic field is generated by currents in the liquid-iron outer core of the planet, and the field changes its strength and direction as the currents slosh around down there. When lava flows cool, little "magnets" in the rock are aligned with the magnetic field and "frozen in." A pile of lava flows of different ages from a volcano thus gives a history of the magnetic field. Such a history is required to read the sun's activity from beryllium-10 in the ice cores—subtract the influence of the magnetic-field changes, and what remains is from the sun.

About 40,000 years ago, the strength of the magnetic field went nearly to zero for much of a millennium. Lava flows record this event, and the large spike in beryllium-10 from the event is seen in sediment piles and ice cores from around the world. But the climate records don't follow the beryllium-10—a huge change in the magnetic field and in the cosmic ray bombardment of Earth produced no noticeable change in climate. We cannot rule out that a small effect occurred, too small to be seen in the records, but any effect was surely not big.

Could anything else from space matter? Micrometeorites also fall to Earth, and if enough of them appeared all at once, they might block the sun and cool the planet. The meteorite that killed the dinosaurs 65 million years ago probably did this. There is no record of this in ice cores, because we don't have any ice that old, but many sediment piles have pieces of that meteorite. Measurements on ice cores show that the rate at which micrometeorites fall has been very small, and essentially constant, as far back as we have looked in the ice, so they do not matter to our story.

As described in the previous chapter, the major way that space affects Earth's climate is through the slow Milankovitch cycles. The stately orbital swings of obliquity, precession, and eccentricity bring on the ice ages and then melt them away over tens of thousands of years. Ice cores and sediment cores show this history beautifully, tracking the cooling as the ice grew and the warming as the ice shrank.

But the ice cores also show us something else very interesting and important. The temperature change from glacial to interglacial was very large on the ice sheets. In central Greenland, it was more than 20°C (36°F). The whole world cooled at the same time as Greenland. Yet the total sunshine reaching Earth was essentially unchanged; it just shifted a bit from pole to equator and from north to south. How is that possible?

As shown in chapter 1, the great reflectivity of the snow and ice helped cause cooling. The ice cores show that the ice age was dustier than today, and the dust blocking the sun would have cooled Earth's surface a little more. Grasslands or tundra replaced some forests during the ice age, and the greater reflectivity of tundra or grassland than forest caused a little more cooling. However, taking all of these effects together is still not sufficient to explain how the whole world changed as much as it did.

The ice cores give us more clues. During the ice age, carbon dioxide was notably reduced, as were methane and nitrous oxide. These are all greenhouse gases, with carbon dioxide the most important factor. If the effect of the decrease in greenhouse gases is included, the ice-age cooling makes sense and is just about the right size, whereas no successful explanation of the cooling has been produced while ignoring the level of greenhouse gases. The ice cores and sediment cores show that the changes in Earth's orbit produced changes in many things, including a little change in temperature, that these led to changes in greenhouse gases, and then the temperature changed more in response to the greenhouse gases, which all makes good sense.

Taken together, the ice-core record and associated studies provide a good overview of Earth's climate system. The sun matters a lot but doesn't change much, volcanoes cause notable but short-lived fluctuations in climate, raising greenhouse gases warms the planet and shrinks ice sheets. But some other things such as changes in dustiness only matter a little, orbits matter a lot but only over long periods of time, and "exotica" such as cosmic rays or the magnetic field matter very little or not at all. But this summary skips the most interesting story from the ice cores…

Dansgaard–Oeschger Events

Although the ice cores show how the climate has responded to changing orbits, energy from the sun, and volcanic eruptions, no one looking at the Greenland records for the first time notices any of this. Instead, one's eye is immediately drawn to the big jumps. A couple of dozen times over the last 100,000 years, a sudden warming occurred over Greenland, often 10°C (18°F) or even more, in roughly a decade or less. Then gradual cooling was followed by fast cooling, a few centuries of really cold conditions, before another warming jump.

These events were first highlighted in papers by the great ice-core scientists Willi Dansgaard and Hans Oeschger, based on the records from the Camp Century and Dye 3 ice cores in Greenland. Wally Broecker named these sharp spikes in the ice record "Dansgaard–Oeschger events."

These events were first discovered in ice very close to bedrock. Both Camp Century and Dye 3 are toward the coast, where the ice is a bit thinner than in the middle of the ice sheet, with fast enough snow accumulation that the ice-age layers have been buried deeply and thinned a lot. When the thinning is very large, one worries that a layer might have pinched off, losing the record of some time and so making a climate change look faster than it really was. When the bed is close, one worries that flow over bumps in the bed might have folded some of the layers, putting ice of different ages close together and simulating a rapid change that never really happened. Dansgaard and Oeschger believed that the similarity of the records from Camp Century and Dye 3 meant that the changes really had occurred, but the worries whether "Dansgaard–Oeschger" events were real or artifacts of the ice flow persisted.

To address this concern, Wally Broecker and others, including Dansgaard and Oeschger, pushed for coring in central Greenland because the ice-age ice would be much farther above the bed there. Two cores in the ice from different locations near the dome would eliminate any sampling errors and demonstrate what had really happened. As it turns out, the central Greenland cores provided the same record back to 110,000 years, and that record matches all other available cores including Camp Century, Dye 3, as well as a newer core at North GRIP, and a core from a small ice dome called Renland in the mountains of east Greenland.

The agreement among the various cores finally offered convincing proof that the abrupt climate changes really happened. Careful inspection of those abrupt changes showed something even more important.

Flipping the Switch

Various indicators, including the isotopic ratios of ice and of trapped gases, documented large and rapid temperature changes in Greenland. The thickness of annual layers was greatly reduced in the cold times compared to the warm times, showing that snowfall was reduced in Greenland during those cold times. So, climate changed in Greenland, repeatedly and rapidly, in patterns that repeated themselves.

The dust and other chemicals trapped in the ice also changed when the climate changed in Greenland, with the cold times being especially dusty. The dust changes were huge, much larger than can be explained by the shift in snowfall altering the dilution of the dust. When snowfall was lower, much more dust was being delivered to the ice sheet. The dust has the chemical "fingerprint" of central Asia, and that fingerprint did not change between warm and cold times. Attempts failed to explain these huge changes in dust based on subtle shifts in wind speed or direction. Instead, these attempts pointed to a larger dust source in Asia during the cold periods in Greenland. Drought favors dust, so the ice cores point to a link between cold in Greenland and windy and dusty times in central Asia.

The gases in the ice cores really drove this point home. Methane has been especially important. As noted above, naturally most methane is produced in wetlands, which especially occur in wet regions of the tropics and in the far north. During the ice age, much of the north was covered with ice. Sea level was lower, because a lot of water that evaporated from the ocean was sitting on land in ice sheets. The lower sea level revealed more land in the tropics, so tropical sources of methane were especially important then. When the temperature fell in Greenland, so did methane, and methane began to recover just as the warming occurred in Greenland. The methane changes were slower than

Figure 2.9
The thin coastal ice of South Greenland, shown here, has been retreating since the end of the Little Ice Age, revealing bedrock scoured by the moving ice, and leaving boulders scattered across the landscape. For scale, note the authors Denton (left) and Conkling in the center of the photo. (Photo by Philip Walsh)

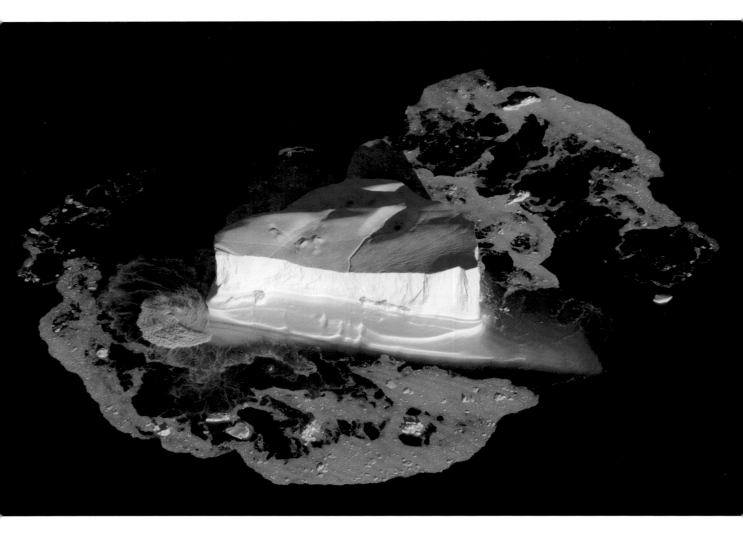

the temperature changes in Greenland, so the greenhouse effect of the methane was not dominating the temperature change.

Methane measurements on ice cores from Antarctica showed changes almost as large as in Greenland, confirming that the tropics were involved. Because some methane is broken down before being mixed uniformly around the world, a methane rise driven solely from the far north would appear notably larger in Greenland cores than in Antarctica, which is not the case.

Nitrous oxide changed with methane, too. Much nitrous oxide is produced with methane in wetlands, but the oceans also make nitrous oxide, complicating interpretations. Nonetheless, the sources of nitrous oxide are widespread, so local effects cannot explain the observed changes. Both the nitrous oxide and the methane histories require that the climate of large parts of the globe changed at the same time that Greenland changed.

Careful studies on Antarctic ice cores, synchronized to the Greenland records using the methane changes, show that atmospheric carbon dioxide underwent subtle shifts at the same time as the Greenland temperature jumps. Since carbon dioxide is not controlled by any single spot on the planet, but by processes acting across vast swaths of Earth's surface and significantly involving the deep circulation of the ocean, this finding was especially suggestive.

Thus, the Greenland ice cores were screaming at us that weird things have repeatedly occurred in Earth's climate system. Like someone flipping a light switch, abrupt warmings and increases in snowfall in Greenland were accompanied by return of rains to central Asia and probably elsewhere on the planet, and shifts in big parts of the oceans as well as the land. These abrupt jumps happened repeatedly. Mere years or a very few decades were involved each time, and the new conditions stuck around for centuries before jumping back.

We humans can "hold on" during a short-lived change caused by a big volcanic eruption, and we can learn the new ways of surviving during the slow change as Earth's orbits initiate ice ages and then end them. Dealing with a big, widespread, abrupt climate change, however, would not be so easy. But the lessons of the Greenland ice cores are very clear: big, widespread, abrupt climate changes have happened repeatedly before there was a large number of "modern" humans on the planet.

Figure 2.10
An iceberg falling apart as it warms in the clear, greenish waters of Scoresby Sound. The debris from small collapses of the iceberg only partially obscures the 90 percent of the iceberg that is below the surface. Notice that the underwater part of the berg is larger than the part sticking up, one of many dangers to a mariner from icebergs and a reason to give them a wide berth. (Photo by Gary Comer)

A ROLE FOR ALL SEASONS

The Discovery of the Younger Dryas

Dryas octapedula, variously known as mountain avens, white dryas, or white dryad is a lovely little arctic-alpine flowering plant usually found growing as a small prostrate evergreen shrub with an eight-petaled white flower. This hardy plant, which grows today in the tundra of the Arctic, has also played a critical role in the study of paleoclimates.

More than a century ago *Dryas* pollen was discovered in abundance in the late-glacial deposits of northwestern Europe, which suggested that Arctic tundra conditions had once blanketed northern Europe. Axel Blytt and Rutger Sernander recognized that alternations in the color of peat layers of late-glacial age recorded switches in climate from warm to cold and back again. From studies of fossil pollen grains in these layers, Lennart von Post and his fellow workers showed that these color changes were the result of oscillations between tundra and boreal environments. The pollen of *Dryas octapedula* occurred in three distinct layers in late-glacial time, each of which was named after the Arctic flower. When translated into English, they were called in ascending order Oldest Dryas, Older Dryas, and Younger Dryas. The Younger Dryas layer was the most striking of the three, and it soon became the hallmark of the

Figure 3.1
Mountain avens or white dryas (*Dryas octopedula*) is the arctic-alpine flowering plant, which migrated far southward during the glacial advances of the Younger Dryas cold reversal. (Photo by Richard Alley)

most recent late-glacial climate reversal. Paleoecologists recognized that the Younger Dryas marked a time when, after the initial postglacial warming and the migration of boreal tree species into northwestern Europe at the end of the last ice age, tundra once again spread southward.

The emergence of the radiocarbon dating method in the 1950s enabled researchers to determine the approximate timing of the Younger Dryas climate reversal. By the middle 1970s Jan Mangerud of the University of Bergen in Norway was able to establish age boundaries on all three Dryas events, with specified ages for the upper and lower boundaries.

At the same time radiocarbon dates began to emerge for a great system of moraines that could be traced through vast tracts of Norway, Sweden, and Finland. These moraines marked a "still-stand," or even a minor readvance, that was superimposed on general recession of the Scandinavian Ice Sheet from its expanded configuration at the last glacial maximum. At the time of deposition of these moraines, the Scandinavian Ice Sheet was about half of its size at the last glacial maximum. The new radiocarbon dates showed that this great system of moraines was the same age as the Younger Dryas tundra layers in late-glacial peat. So it became clear that the signature of the Younger Dryas climate event was registered in at least two archives—one biological and one geological. But all of this preliminary work was trumped when the classic signature of the Younger Dryas climate event was discovered in long cores through the Greenland Ice Sheet.

The Younger Dryas in the Ice Core Record

As described in chapter 2, deep ice cores began to be collected from the Greenland Ice Sheet in the 1960s. The first of these cores was taken at Camp Century, located on the ice sheet in northwest Greenland 100 miles inland from the coast. The oxygen-isotope record taken along the length of the ice core from Camp Century showed the Younger Dryas climate reversal in considerable detail. Another ice core, taken a few years later at Dye 3 in south-central Greenland, displayed the same detailed signature of the Younger Dryas. Then the GRIP and GISP2 cores were drilled at a site called Summit on the main interior divide of the ice sheet. Within the last few years, at a drill site dubbed North GRIP, or NGRIP, located north of Summit on the inland divide of the ice sheet, a European team obtained an even newer core. All of these cores showed the same Younger Dryas climate signals as those first discovered at Camp Century.

One of the most important discoveries from the network of ice cores is that they all showed that the Younger Dryas climate signal was pervasive over the entire island of Greenland. In particular, this last core, NGRIP from north-central Greenland, provided a particularly accurate chronology for the Younger Dryas signal. This chronology was developed by the counting of annual layers in the ice core and demonstrated that the onset of the Younger Dryas occurred 12,900 years ago, and its demise 11,700 years ago. Thus the total length of the Younger Dryas cold reversal superimposed on the last deglaciation was 1,200 years. A stunning discovery from these early ice cores was the suddenness with which the Younger Dryas began and ended, in each case a matter of a few years. This was the first indication that the last glaciation, at least in the North Atlantic region, was marked by abrupt climate changes.

Another important discovery recorded in all of the Greenland ice cores was that the Younger Dryas was not unique. Rather it was simply the latest in a series of 24 or so very similar climate reversals that punctuated the last glacial cycle. Wally Broecker named all of these older oscillations Dansgaard–Oeschger events in recognition of two of the early workers who studied these abrupt climate changes. Willi Dansgaard of the University of Copenhagen in Denmark had established the oxygen-isotope records of the Camp Century and Dye 3 ice cores, while Hans Oeschger, a physicist at the University of Bern in Switzerland, had studied the gases preserved in bubbles in the ice cores. Along with Chester Langway of the Cold Regions Research and Engineering Laboratory in Hanover, New Hampshire, they discovered many older events similar to the Younger Dryas, which showed that abrupt climate changes were a common feature of the last glacial cycle. These abrupt climate changes demanded an explanation that we did not have.

Intimations from Earlier Expeditions

George Denton's first encounter with Younger Dryas moraines took place in early summer of 1966. He had just finished a winter's work of cutting ice out of the snout of Storglaciaren in northern Sweden using a chain saw equipped with an appropriate ice blade. This small valley glacier drained eastward from Kebnekaise, the highest mountain in Swedish Lapland, situated about 150 kilometers north of the Arctic Circle. Once sawed out of the glacier in cubic blocks, the ice was transported back to the nearby Tarfala mountain station, either by ski-dog-drawn sled or by dog team. There it was loaded into a huge metal tank, placed under vacuum, and heated. The air released from the air bubbles in the ice was passed through a molecular sieve, also under vacuum, to collect the CO_2 from this air. The sieve with the extracted CO_2 was transported to Bern, where Hans Oescher determined the radiocarbon age of the CO_2, and therefore the ice in the snout of Storglacieren.

Work on the Stornglacieren ice began long before the advent of accelerator radiocarbon dating. So to obtain the age of CO_2 from bubbles in glacier ice, Hans had painstakingly built tiny counters to measure the small gas samples of CO_2 collected from the enclosed bubbles. All of this was very tedious work and took most of the time available in the course of the winter months when they had collected the ice. The only break came in early spring when full light returned after the long Arctic night. Lasse Sarri, a Lapp from the nearby village of Nikkaluokta, took the members of the ice-dating team on a week-long ski trip through the spectacular mountains of Swedish Lapland, where we followed the famous Kungsladan trail between Abisko and Kebnekaise and spent nights in Lappish huts near the trail.

Once the field season in Swedish Lapland was finished early in the summer of 1966, Denton took a working holiday to visit Professor Bjorn Andersen, newly installed at the University of Oslo. At the time, Andersen was working in southwestern Norway, mapping the distribution of moraines near his boyhood hometown of Stavanger on the coast. Together Andersen and Denton traveled a short distance north of Stavanger to view the famous Esmark Moraine that had been deposited during the last ice age by a glacier in Lysefjord.

In 1824 the Norwegian geologist, Professor Jan Esmark, had recognized that this moraine, a classic loop ridge with a relief of 20 to 30 meters, represented the

Figure 3.2
"Late-glacial" moraines formed during the "bumpy" warming from the last ice age, including during the Younger Dryas, appear as sinuous parallel ridges on the sloping edges of this glacial valley in the mountains surrounding Scoresby Sound. Little Ice Age moraines are gray and are visible at the head of the melt water stream, and the still-flowing glaciers end farther up the valley. (Photo by Philip Conkling)

snout of a former glacier. He came to this conclusion by comparing the Esmark Moraine with similar ridges at the fronts of glaciers in the high mountains of Jotenheimen in south-central Norway. In the same year Esmark was the first to postulate a widespread glaciation in northern Europe. We now know that the Esmark Moraine was deposited in Lysefjord by a tongue of the Scandinavian Ice Sheet during the Younger Dryas cold reversal.

Andersen and Denton then followed the Younger Dryas moraine belt up onto a plateau between fjords. Here a single Younger Dryas moraine was only a few meters high and had the appearance of a New England stone wall, which according to an ancient legend had been built by a troll, as giants were known, to keep neighboring trolls out of his territory. And so the moraine was known as Trollgaren, or giant fence.

After examining the fence built by this giant, Andersen and Denton traveled to the moraines of Younger Dryas age in the Oslo fjord. There the outermost of these moraines made up an extensive complex called the Ra, which was deposited by a glacier that pushed southward through the Oslo fjord. Andersen determined the radiocarbon age of this and other such moraines from the dates of shells in marine deposits, such as deltas, linked to the ridge itself. It turns out that the Ra ridges were deposited near the beginning of Younger Dryas time. But the Ra ridges were not the only moraine ridges of Younger Dryas age in the Oslo fjord region. There were two additional sets of moraines farther up the fjord that were also deposited during Younger Dryas time. Thus, the Younger Dryas in the Oslo fjord region was marked by overall glacier retreat of about 40 kilometers that was interrupted three times by short-lived pulses resulting in the deposition of the three moraine complexes.

Much later, in 1995, Denton accompanied a group of Scandinavian glacial geologists on a field excursion along the tier of Younger Dryas moraines in southern Norway, southern Sweden, and southern Finland designed to summarize the glacial history of the last ice-age cycle. In Norway the Younger Dryas moraine complexes in the Oslo fjord region were the Ra, the As, and the Ski; in Sweden they were the Skovde and the Billingen; and in Finland they were the Salpausselka I, II, and III.

Just as in the Oslo fjord, the distribution of moraines in Finland, and to a minor extent in Sweden, suggests overall recession of the southern margin of the Scandinavian Ice Sheet in the course of the Younger Dryas, with halts or

Figure 3.3
A spectacular late-glacial moraine loop in Itivdlerssuaq Valley in southern Greenland. The band of moraines is seen most easily in the lower left, between the lake and the stream. A small glacier (not visible, to the upper left) occupied the entire loop during late-glacial times, and extended into the picture during the Little Ice Age to leave the non-vegetated region in the valley bottom toward the upper left, as shown in figure 8-2. (Photo by Gary Comer)

minor readvances that resulted in moraine deposition. In the case of Oslo fjord, the overall recession in Younger Dryas time was about 40 kilometers, while in southern Finland it was tens of kilometers.

In addition to the field trips to the Younger Dryas moraines in Scandinavia, Denton also visited the moraines of the same age in the Swiss Alps with Christian Schulchter and Max Maisch, as well as visiting moraines in the Scottish mountains with David Sugden and Chalmers Clapperton. In the case of the Swiss Alps, the moraines of Younger Dryas age were dubbed Egesen. These moraines were situated deep in the alpine valleys not far distant from the Little Ice Age moraines of the mid nineteenth century. Max Maisch calculated that the snowline drop below the Little Ice Age value for the Egesen moraines was only about 300 meters. In contrast, the snow line drop for the maximum of the last ice age was 1,200 meters, nearly four times as much as for the Egesen moraines. In western Scotland the Loch Lomond moraines were deposited during the Younger Dryas. These moraines were situated well within the ice limits of the last glacial maximum. In eastern Scotland, the moraines of Younger Dryas age were restricted to cirques, one of which was situated on the British royal estate of Balmoral.

The major problem that began to emerge from the examination of the Younger Dryas moraines in Europe during these sporadic visits over the course of nearly four decades was that the moraines pointed to a rather modest climate reversal during Younger Dryas time. However, in sharp contrast, reports from the Greenland ice cores suggested that Younger Dryas climate reverted nearly to full-glacial conditions. Something was amiss, but it was hard to put one's finger on exactly where the problem lay.

The situation became even more confusing after Hans Oeschger obtained dates developed from oxygen-isotope records for a number of marl lakes near Bern in Switzerland. Oeschger's results demonstrated that the oxygen-isotope signature of the marl lakes was a dead ringer for the signature in Greenland ice cores. The isotopic signature in marls in Swiss lakes showed a Younger Dryas climate reversal back toward the values of the last glacial maximum. We could offer no explanation for the major temperature mismatches detected by the moraine and the isotopic records.

The confusion increased when Jeff Severinghaus, using clever isotopic techniques, showed that the mean annual temperature at Summit in Greenland,

Figure 3.4
Little Ice Age valley outlined in gray moraine deposits in Jameson Land of Scoresby Sound with the Stauning Alps in background. Note the parallel sinuous ridges of late-glacial moraines including the Younger Dryas in the lower-left foreground. (Photo by Philip Conkling)

where the GISP ice core was located, plunged to 16°C below today's values during the Younger Dryas cold snap. The snowline drop associated with moraines in Europe was far in excess of that implied by this sharp temperature decline. Why the difference? The answer was not obvious, and it was convenient to sweep the problem under the rug by attributing it to some peculiarity of Greenland climate.

The Lake Agassiz Flood Theory

Then along came Gary Comer, who had become interested in climate change science following his voyage through the legendary Northwest Passage. After talking to Wally Broecker, Gary decided that the first field trip he wanted to take was to western Canada to view the famous channels that were the focus of the great flood theory that postulated a deluge through the Great Lakes, down the St. Lawrence Seaway, and into the North Atlantic from Glacial Lake Agassiz. Lake Agassiz was the largest proglacial lake to form in northern North America during the last glaciation. It was dammed against the southwest margin of the Laurentide Ice Sheet during its retreat from the maximum position of the last ice age.

One of the most compelling ideas in the scientific literature had been that floods from Lake Agassiz had changed the buoyancy of the North Atlantic and curtailed overturning circulation of the northern limb of the ocean conveyor current, thus triggering the Younger Dryas cold pulse (see chapter 4). The fact that the Younger Dryas was only the latest in a long line of 24 such cold reversals during the last ice age suggested that there had been at least two dozen such floods.

A group of field geologists met at Gary's farm in Wisconsin to plan the trip. They laid out flight paths over the flood channel of presumed Younger Dryas age in southern Canada, and discussed the best strategy to examine and date the evidence of the flood. Once organized, the group flew north to Thunder Bay on the northern shore of Lake Superior in southern Ontario. A Caravan fixed-wing aircraft and a helicopter were waiting. One group boarded the helicopter to examine channels cut into bedrock north of Thunder Bay. They believed these channels postdated the Younger Dryas and therefore were not of prime interest. The rest of the group boarded the Caravan for the regions beyond Thunder Bay. Their flight path took us over the postulated position of the prime channel through which they believed floodwaters from Lake Agassiz passed to Thunder Bay and eastward through the Great Lakes to the Saint Lawrence Seaway and the North Atlantic.

They flew for two days looking for this channel, but it was nowhere to be seen. They found this situation to be extremely puzzling. A whole cottage industry had been set up around the idea that a flood passing through this supposed channel had triggered the Younger Dryas by shutting down the overturning conveyor circulation in the North Atlantic. But such a trigger obviously did not act through the postulated channel; there was no such channel.

To check the situation out carefully in subsequent years, they implemented a more thorough investigation of all the potential channels for a Lake Agassiz flood into the North Atlantic and even the Arctic Ocean. They delegated field teams under the direction of Tom Lowell and Tim Fisher to map the channels and to date the nearby moraines. The field and chronologic studies were carried out in the Thunder Bay area and in the southern reaches of the former Glacial Lake Agassiz. In addition, the same field teams were dispatched to the vicinity of Fort McMurray in northern Alberta to carry out similar studies of a postulated former flood to the Arctic Ocean. The results were very helpful in determining the chronology of recession of western margins of the Laurentide Ice Sheet. But they failed to identify an appropriate channel of the correct age to tie into the beginning of the Younger Dryas event. The Lake Agassiz flood theory was in trouble.

The Scoresby Sound Expedition

One day during the trip to Thunder Bay, Gary Comer and George Denton had lunch together. Gary asked whether George would be interested in traveling to Greenland with him on the M/V *Turmoil* during the coming summer to examine

Figure 3.5
Entrance to Scoresby Sound with Cape Brewster to the left. (Photo by Philip Conkling)

Figure 3.6
Aerial view of the ice cap near the entrance to Scoresby Sound looking south toward
the Renland Plateau. Flow of glaciers over bumps in their bedrock generates many of the
spectacular crevasses. (Photo by Philip Conkling)

further the problem of the Younger Dryas. And if so, where should they go? George's answer was that he would certainly go and that Scoresby Sound in East Greenland was the place they should head.

Why Scoresby Sound? The answer involved the puzzle about the mismatch between moraine and ice-core records that Denton had encountered earlier when examining the moraines of Younger Dryas age in Europe. Perhaps it could be resolved in Scoresby Sound. This was the only place in the Northern Hemisphere where an ice core and moraines occurred side by side.

Several decades ago the Danes had recovered an ice core from the Renland plateau ice cap near the head of Scoresby Sound. Renland itself was a flat-topped block of bedrock that had been isolated off the eastern periphery of the Greenland Ice Sheet by headward erosion of deep fjords. Its top was covered by a thin plateau ice cap that fed outlet glaciers spilling down its sides and into the adjacent fjords. Despite the fact that the ice thickness at the drill site was only 360 meters, the Renland core showed all of the major climate features of the inland ice cores of the GISP2 and GRIP projects. The Younger Dryas signature was particularly prominent.

More than four decades ago, Svend Funder, a well-known Danish geologist, began a long-term study of the glacial history of eastern Greenland, with a focus on the Scoresby Sound region. Of greatest interest to us was his discovery and mapping of a prominent moraine system in the inner reaches of Scoresby Sound. Funder referred to this system as the Milne Land moraines, named for the time in which they were deposited the Milne Land Stadial. These moraines were deposited by a coalescent composite of glaciers, including outlets of the main ice sheet, as well as alpine glaciers from the Stauning Alps and outlets from the plateau glacier on the island of Milne Land.

Most important, outlet glaciers from the plateau ice cap on Renland also fed the composite glacier that deposited the Milne Land moraines. When the glaciers that had deposited the Milne Land moraines retreated at the end of the Milne Land Stadial, the land that had been pushed down by the weight of the ice began to rebound. This rebound was so rapid that its pace for a time exceeded the rise of sea level that accompanied the dissipation of ice sheets during the end of the last ice age. The result was the upheaval of marine deposits and deltas to positions well above even present-day sea level. Funder could determine the timing of this uplift from radiocarbon dating of the shells enclosed in the

Figure 3.7
George Denton (right) and Bjorn Andersen study the geomorphology of a raised delta colonized by tundra plants, including bearberry leaves that show this striking red color in autumn, in Milne Land along the northern flank of Scoresby Sound. (Photo by Richard Alley)

Figure 3.8
Village of Scoresby on the
northern shore near the
entrance to Scoresby Sound.
(Photo by Philip Walsh)

Figure 3.9
Retreating alpine glaciers in
the Stauning Alps. Note the
jagged peaks on the mountain
spires that are above the "trim
line" limit of ice cover during
glacial maximum. Flow of
ice over the bedrock rapidly
removes such relatively
breakable features, producing
the smoother appearance
of bedrock in the lower part
of the photo. (Photo by Philip
Conkling)

raised marine sediments. The results suggested to him that the outermost Milne
Land moraines dated to Younger Dryas time.

So the situation was perfect. Side by side stood an ice core with the Younger
Dryas signature and a moraine set of Younger Dryas age. With two independently
dated climate records so close to one another, perhaps Scoresby Sound
would hold the answer to the scientific mystery of the huge difference in temperatures
in Greenland and northern Europe during the Younger Dryas.

On a beautiful August day in Maine, Gary Comer and George Denton, joined
by Richard Alley and Philip Conkling, flew out of the Hancock County airport
near Acadia National Park, bound for Reykjavik, Iceland, where M/V *Turmoil*
was awaiting their arrival. About three and a half hours later, they landed in
Iceland and were whisked to M/V *Turmoil* in Reykjavik Harbor. The next morning
they took an aerial tour over some of the Icelandic ice caps, and then, back
on board the *Turmoil*, left Reykjavik Harbor for an anchorage in a beautiful
fjord in the northwestern sector of the island. After another aerial tour of the
local Dranga ice cap, they were underway for Scoresby Sound, about 36 hours
cruising time from northwestern Iceland.

Figure 3.10
Upper Qinguadalen Valley near Julianehab in southern Greenland. Note that the upper reaches of these glaciers are snow covered, while the lower reaches are bare ice. The highest elevation of the boundary between snow and bare ice in late summer approximates the equilibrium line of the glacier separating the upper accumulation zone from the lower ablation zone. Reconstructions of glacier outlines are used to estimate the locations of former snowlines, which are important indicators of past climates, rising with warming and falling with cooling. Marginal moraines, where they form, are restricted to regions below the equilibrium line, so upper limits of moraines can be used to reconstruct former snowlines. (Photo by Gary Comer)

After a rough crossing, they passed through the entrance of Scoresby Sound and set anchor in the harbor alongside the village of Scoresby. The fixed-wing Caravan aircraft that had flown them over the ice caps of Iceland was equipped with pontoons and therefore could land in the harbor beside us. They climbed aboard and flew inland toward the Stauning Alps, while the *Turmoil* weighed anchor and cruised toward the head of the sound to a harbor in the Bear Islands near the mouth of Northwest Fjord.

The likely answer to the long-standing puzzle of the difference in ice-core and moraine climate signatures was immediately obvious as soon as they flew near the Stauning Alps and viewed the Milne Land moraines and the Renland ice cap in one vista from the plane window. They could do so because the drill site was located only 50 kilometers from the moraines. If the Milne Land moraines were truly of Younger Dryas age, the reason the climate records showed such a mismatch had to be the markedly different summer and winter seasonal temperatures of Younger Dryas climate. Seasonality was the key to solving the puzzle.

To grasp the solution, it is important to know that glaciers respond predominately to changes in summer temperature. The magnitude of summer-temperature lowering responsible for a given advance can be determined from the moraines deposited at the culmination of the expansion. Of most interest is the extension of the lateral moraines uphill alongside the glacier margins. The uppermost two-thirds of a glacier is the accumulation zone, where snowfall exceeds melting. Snow and ice, and rocks, move into the glacier in the accumulation zone, so moraines are not deposited around it. The lower third of the glacier is the ablation zone, where melting exceeds snowfall, and rocks are carried to the glacier edge around the ablation zone and left there as the ice melts, forming lateral moraines. The line dividing the accumulation and ablation zones is the equilibrium line, and its elevation moves up with warming and down with cooling. In late summer, the equilibrium line is very close to the snowline, the lowest elevation with snow cover. Many field workers refer to the snowline for simplicity, because it is easy to say and to locate in late summer, and that will be done here.

During advances and retreats of glacier termini, the snowline moves up and down in altitude by an amount determined by the lapse rate in the local atmosphere. By tracing the lateral moraines upslope to their highest position

alongside the glacier margins, one can determine the altitude at which they die out. This altitude will be that of the former snowline at the time when the moraines were deposited. By comparing this paleoaltitude with the present snowline of the glacier in question, and making use of the local atmospheric lapse rate, one can calculate the summer temperature lowering that was responsible for the advance of the glacier tongue to the moraines in question.

When these principles are applied to the Milne Land moraines, a consistent value of 3–6°C for summer temperature decline as compared to today is obtained. The value of 16°C lowering of temperature below today's values that Jeff Severinghaus had measured in the GISP2 ice core, some 200 kilometers inland from the head of Northwest Fjord, represented *mean* annual temperature. If the summer temperature declined only 3–6°C and the mean annual temperature was 16°C below today's values, then there can be only one conclusion. The climate must have been highly seasonal, with winter temperatures plunging below today's values by 26–28°C. This meant that the climate over the North Atlantic during the Younger Dryas was akin to that of Siberia today.

Figure 3.11
Sea ice over the North Atlantic caps the transfer of heat from ocean to atmosphere, causing regional temperatures to plunge across the Northern Hemisphere. Here, open-water "leads" snake through sea ice, formed by freezing of ocean water and flow past icebergs formed by snowfall on glaciers. (Photo by Gary Comer)

The Sea Ice Hypothesis

But how can that be? The team quickly realized that the mechanism necessary to produce such Siberian climate conditions must have been produced by the spread of a winter sea-ice cover over the northern North Atlantic Ocean. This would have been akin to putting a cover on a swimming pool during hours of darkness. The transfer of heat from the ocean to the atmosphere would have been curtailed, allowing the temperature of the overlying atmosphere to plunge. The lowering of the summertime temperature, recorded by the glaciers, would have been much less because of the melt back of the ice cover. Of course, the reason the sea-ice cover was able to spread so dramatically was that the overturning ocean circulation in the North Atlantic was curtailed during Younger Dryas time.

The team was delighted to learn that Dutch scientist Hans Renssen and his colleagues working in Europe had also reached similar conclusions, if for quite different reasons. They examined the seasonality of the Younger Dryas climate

in northwest Europe alongside the North Atlantic by establishing the distribution of so-called indicator plants and of periglacial features associated with frozen ground. The indicator plants gave a summer temperature lowering of about 3–6°C compared to today's values. In sharp contrast, the frozen ground indicators gave a lowering of mean annual temperature of about 12–16°C. So once again a winter lowering of about 26°C was necessary to achieve the annual value. These numbers are close to the values obtained in the Scoresby Sound region of East Greenland by comparing mountain glaciers and ice cores. So the seasonality signature, marked by severe winters, was widespread in the North Atlantic region.

Within the last two years another important contribution to this problem has emerged from a study of Lake Meerfelder Maar in western Germany by Achim Brauer and colleagues. The sediments deposited in the basin of this lake show annual layers. So the rapidity of the onset and termination of the Younger Dryas climate changes can be ascertained with great accuracy by counting the layers. What the record shows is astounding. The regular, repetitive bedding of the sediments was suddenly replaced at the onset of the Younger Dryas by reworked winter layers produced by strong winds blowing eastward off of the North Atlantic. These winds were cold because of a winter sea-ice cover that extended southward to latitude 50°N. The winds represented a shift to zonal flow of the westerlies, with much stronger winter winds than during the previous warm interval. The most startling aspect of the record is that this sudden change in sedimentation on the lake floor took place between one year and the next. The indication is that at the latitude of Meerfelder Maar, the onset of the Younger Dryas, marked by a change in wintertime wind strength, was particularly sudden and was associated with the spread of sea ice across the North Atlantic.

One might ask why a cold North Atlantic with widespread sea ice, particularly in winter, is so important for the understanding of abrupt climate change. Before addressing this question, it should be noted that abrupt climate changes with timing similar to that in Greenland are widespread elsewhere in the Northern Hemisphere and in the tropics. For example, sudden changes with the same signature as those in Greenland are evident in records of the Asian monsoon and of changes in currents off of Santa Barbara in California. The Intertropical Convergence Zone, which marks the thermal equator of the earth, also shows north–south switches that follow the millennial-scale changes in Greenland. So any explanation of the abrupt climate changes recorded in Greenland ice must take into account the widespread nature of abrupt climate change. This is where the North Atlantic sea ice comes into play.

Seasonality in the North Atlantic turns out to be important because model results suggest that the expansion of sea ice across the northern North Atlantic, particularly in winter, is the key factor in spreading the impacts of the millennial-scale cold events quickly and efficiently throughout the Northern Hemisphere and into the tropics, as discussed in chapter 6.

In summary, this notion of highly seasonal climate in the northern North Atlantic during the cold phases between abrupt warming periods followed by cold reversals recorded in Greenland ice cores helps to explain how the Greenland climate is propagated globally. These discoveries also raise the question of how any warming phase affecting Greenland's climate might also be projected globally.

4

THE GREAT OCEAN CONVEYOR

Early Research in the Deep Ocean

The *conveyor* is one of the ocean's great global current loops. It originates in the northernmost regions of the Atlantic Ocean where, during the winter, frigid air flowing off Canada and Greenland cools, and hence densifies, the salty waters carried into this region by the Gulf Stream. The result is that the surface water becomes dense enough to sink into the abyss to form what is known by oceanographers as North Atlantic Deep Water. This water drifts southward down the length of the Atlantic. When it reaches the tip of Africa, it joins the ocean's mighty Mixmaster, a circular torrent driven round and round the Antarctic continent by the force of the southern ocean's westerly winds.

Balancing the Atlantic's inflow to this torrent (and also the input of deep water generated along the margins of the Antarctic continent) are outflows into the abyssal Indian and Pacific Oceans. The ocean's global current loop is completed when these waters well up to the surface and flow back to the Atlantic. The part of the upwelling that occurs in the Pacific is largely channeled through the Indonesian Straits into the Indian Ocean and from there around the tip of Africa back into the Atlantic (see figure 4.2).

The term *conveyor* refers to the Atlantic Ocean portion of this vast system. Surface waters, including those of the Gulf Stream, carry tropical heat to the Atlantic's northern reaches where during the winter heat is sucked out of the water and into the atmosphere. So, just as conveyor belts carry coal to furnaces of electrical power plants, the Great Ocean Conveyor delivers heat to the northern reaches of the Atlantic. Physical oceanographers refer to this flow in more formal terms, that is, as the Atlantic's Meridional Overturning Circulation.

Wallace Broecker was initially introduced to the subject of ocean circulation in 1952 between his junior and senior years in college when he served as a summer intern in the radiocarbon dating laboratory at Columbia University's Lamont Geological Observatory (now the Lamont–Doherty Earth Observatory). He learned that his boss, Professor J. Lawrence Kulp, was cooperating with Lamont's director, Maurice Ewing, to determine the rate at which new deep water was being produced in the northern Atlantic in order to estimate the residence time of various water in the deep ocean. The idea was to determine the deficiency of ^{14}C in the dissolved bicarbonate of water from the deep Atlantic relative to its concentration in atmospheric CO_2. As radiocarbon de-

Figure 4.1
Wallace Broecker at Julianehab Icefield. (Photo by Philip Conkling)

Figure 4.2
The Great Ocean Conveyor. (Illustration by Joe Le Monnier, *Natural History Magazine*)

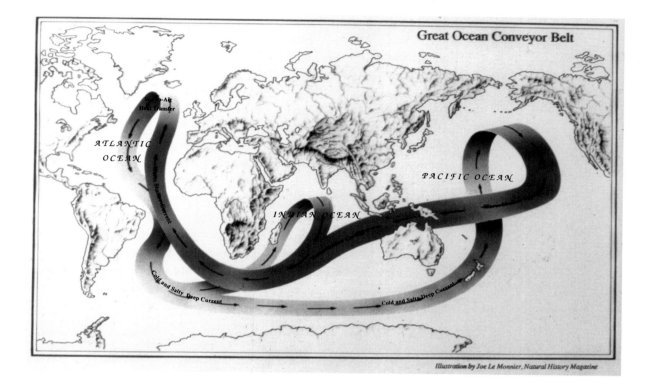

Illustration by Joe Le Monnier, *Natural History Magazine*

cays at the rate of 1 percent in 82 years, quantification of this deficiency would allow them to estimate how long the water in the deep Atlantic had been isolated from the atmosphere. Assuming that the ocean had operated in more or less the same manner over the last thousand or so years, they could convert this isolation time to a deep-water production rate, which was vital to an understanding of global ocean circulation.

Ewing's contribution to this enterprise was the use of Lamont's research vessel, *Vema*, to obtain the 50-gallon samples of water required for each measurement. Initially Ewing employed an oil drum fitted with a ship's porthole to obtain these samples. He sent the barrel down into the deep ocean with the porthole open, allowing water to flush through the barrel. Once it reached the desired depth, the barrel was run up and down to ensure that it was well flushed with the target water. Then a "messenger" was sent down the cable to trigger a release, snapping the porthole shut, after which the sample was hoisted back to the ship's deck.

Rather than shipping these large volumes of water back to Lamont, Ewing and his team extracted the carbon at sea. Kulp provided the required chemicals—acid to convert the HCO_3^- to CO_2 and ascarite to absorb the CO_2 that formed in the chemical reaction. (Ascarite is the commercial name for barium hydroxide absorbed on clay minerals.) Air was then circulated through the acidified seawater until all the CO_2 was captured on the ascarite. Back in the lab, Kulp removed the CO_2 from the ascarite and converted it to the solid carbon black used for ^{14}C measurement. Kulp's early results suggested that the isolation time for deep Atlantic water was on the order of a millennium.

It turned out that a problem with the measurement apparatus created an opportunity that launched Broecker's career. Electrical arcing in the high-voltage pass troughs at the ends of the counting chamber was spoiling the measurement. At Broecker's suggestion, they replaced the glass tubing used as insulators with pieces of Teflon that he fashioned from high-voltage wire that he happened to spot in their small electronics shop. It worked! In Kulp's mind, Broecker had saved the summer by allowing him to get desperately needed data. In this way, by the end of his summer internship Broecker had made a sufficiently favorable impression that Kulp asked him to stay on at Columbia as a senior transfer student. He did, and 58 years later he is still there!

Two years later, in 1954, on the completion of a new laboratory building to house Kulp's rapidly growing geochemistry group, Broecker was put in charge of replacing the solid carbon method to measure ^{14}C's radioactive decay with a new method that involved using CO_2 as the gas in the measuring device. The reason for this switch was that radioactive fission products released to the atmosphere from the above-ground A-bomb tests conducted at the Nevada test site were beginning to interfere with the results. Some of this radioactivity remained airborne all the way across country and found its way onto their carbon black as it was dried. The radioactive emissions by these fission products added to those from the sample's radiocarbon, leading to anomalously young apparent ages.

One of the first things Broecker did when this new radiocarbon facility was completed was to check a key assumption made by Kulp. Although he had a small aliquot of the ascarite from each bottle analyzed for CO_2 before it was sent to sea, he had never measured the ^{14}C content of this CO_2, but rather assumed that ascarite CO_2 had a ^{14}C to C ratio equal to that in atmospheric CO_2. As Ewing's seawater samples had ^{14}C to C ratios lower than those for the atmospheric CO_2, Kulp's correction *increased* the apparent isolation time he calculated for the seawater CO_2.

In order to check Kulp's assumption, Broecker took ten bottles of fresh ascarite and removed the CO_2. The result showed that the ascarite CO_2 contained *no* ^{14}C. When corrected in this way, the apparent isolation time for the deep Atlantic seawater samples became much smaller—closer to a century than a millennium.

As part of his PhD research, Broecker conducted measurements on many of the *Vema* samples. Years later, hundreds of measurements made as part of the GEOSECS program by labs at the University of Miami and the University of Washington not only confirmed the early Lamont results, but showed that the range of isolation times for the North Atlantic Deep Water mass lay mainly in the range of 100 to 200 years.

Ocean Research and Paleoclimate Records

It was not until the mid-1980s that the importance of this early oceanographic research to the understanding of paleoclimate became clear. The reality hit Broecker during a lecture given by Hans Oeschger at Switzerland's Bern University on the results of measurements on the newly drilled southern Greenland Dye 3 ice core. Not only did Oeschger's new record beautifully replicate the millennial-duration sharp ups and downs in oxygen isotope ratio found more than a decade earlier in the northern Greenland Camp Century ice core (see figure 4.3), but his ice-core data showed that these sudden shifts were accompanied by 50 parts per million ups and downs of the concentration of CO_2 in the air trapped in the ice.

Broecker was stunned. During the intervening years, he had spent much effort trying to explain why only about 35 percent of the CO_2 generated by the burning of fossil fuels was being taken up by the ocean. Considering this sluggishness of the

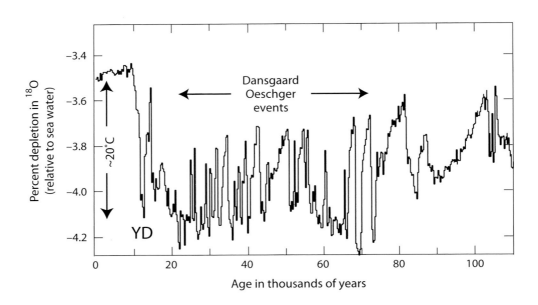

Figure 4.3
Dansgaard's temperature record from the GRIP ice core showing abrupt climate changes. Note that today is on the left, and 100,000 years ago on the right.

Figure 4.4
Field research team, Richard Alley (left), George Denton, and Wallace Broecker (right) in southern Greenland. (Photo by Philip Conkling)

ocean taking up CO_2, he quickly realized the seeming impossibility of transferring the 100 or so gigatons of carbon into and out of the ocean on the time scales of a few decades as required by Oeschger's ice core results. While pondering how this might have happened, he focused his thinking on the northern Atlantic Ocean. It occurred to him that if the conveyor came to a halt, so would its delivery of heat to the atmosphere. So perhaps the conveyor turning on and off could explain the large and extremely rapid temperature changes recorded by oxygen isotopes in ice from both ends of Greenland.

Broecker pondered whether the disruption in the ocean's carbon cycle caused by these conveyor starts and stops could possibly explain Oeschger's large CO_2 changes. But try as he might he could see no feasible way for these jumps to have happened. This conundrum bothered him, but he remained confident that he was on the right track. Fortunately, the question regarding the big jumps in CO_2 in the atmosphere was answered a couple of years later when measurements made by Oeschger's group showed that the abrupt CO_2 increases were artifacts of the reaction between $CaCO_3$ dust and sulfuric acid within the ice, as described in the previous chapter. Furthermore, a detailed CO_2 record from

an Antarctic ice core failed to reproduce these 50 parts per million jumps; we now know that only subtle shifts in CO_2 occurred with these events. So the burden of explanation was off his back! However, had Oeschger not presented his original CO_2 results, he might not have hit upon the conveyor concept. Oeschger's CO_2 jumps caused Broecker to ponder how the ocean might be involved in changing atmospheric CO_2 concentrations.

It is here that the importance of the earlier work with century-isolation time came into play. Broecker's idea required that the shutdown of the heat released from warm waters carried by the Gulf Stream into the Norwegian Sea was responsible for the temperature changes recorded in Greenland. If Kulp's initial millennial time scale proved correct, this heat would have amounted to only a minuscule 3 percent of that supplied to this region by the sun. But using a century timescale, it became a whopping 30 percent.

Another Twist in the Ocean Conveyor Story

The demise of Oeschger's CO_2 story was only the first of a series of many twists and turns taken by the ocean conveyor hypothesis. The next one had to do with the geographic extent of the impacts of the conveyor's numerous stops and starts. Broecker's initial idea was that these impacts were restricted to the landmass downwind of the northern Atlantic. Model simulations showed that the heat released to the atmosphere by the conveyor is lost to outer space on the timescale of a week or two. This being the case, the westerly winds would have been unable to carry heat eastward even as far as Asia, let alone around the globe to North America. This view was supported by the then-available observations of the impacts of the Younger Dryas, the last of the series of Greenland's cold events. Evidence of the Younger Dryas cold advance shows up as a glacial readvance and as a return to cold conditions in sediments. Although well recorded across northern Europe, evidence of Younger Dryas impacts appeared to be absent elsewhere.

This apple cart was soon upset by observations by Broecker's colleague, Dorothy Peteet, who specializes in studies of the pollen record preserved in peat. One day in 1983 she announced that she had found evidence for the Younger

Dryas in a bog only one kilometer away from the entrance to the Lamont property. Broecker hemmed and hawed and commented that it was only a small feature in her record. Undeterred, she extended her search and to Broecker's surprise came up with convincing evidence for a Younger Dryas impact in southern Alaska. Faced with this new evidence, Broecker could no longer call on a shutdown of heat released by the conveyor to explain the intense cold episodes of glacial time. Something much more grand must have occurred.

More Climate Change Evidence from Ocean Sediments

While Broecker was pondering Peteet's findings, he happened to visit Jim Kennett, a marine geologist at the University of California, Santa Barbara. During breakfast on the terrace in front of his home, Kennett pointed to the ocean and said that he had lined up the *Glomar Challenger* to core the sediments "right out there," meaning in the Santa Barbara channel, which lay between the coast and a series of off-shore islands. Beneath this channel was a silled basin whose bottom waters were almost free of dissolved oxygen.

Wally shrugged at Jim's disclosure, never for a moment realizing that his sediment core would revolutionize his thinking, for it turned out to contain a record that beautifully replicated every one of Greenland's Dansgaard–Oeschger events, as well as the Younger Dryas event. Even more amazing was that this discovery didn't even involve the kind of expensive isotope ratio measurements made on Greenland ice. Rather, it was done with the naked eye.

Kennett's sediment core consisted of a series of discrete layers that alternated between those that were finely laminated and those that were free of laminations. Radiocarbon dates provided a chronology, which allowed Kennett to compare his record with that from Greenland ice. It turned out that each of his laminated sections matched in time with one of Greenland's very cold periods (which, in scientific shorthand, we refer to as "^{18}O stadials") and each of his unlaminated sections matched one of Greenland's warmer, but still moderately cold intervals—or "^{18}O interstadials."

The last pair of these alternating layers corresponded to the warm Bølling–Allerød interstadial and the cold Younger Dryas stadial. Kennett's interpretation

Figure 4.5
Hartmut Heinrich first proposed that layers of rock fragments found in deep-sea sediment cores drilled in the North Atlantic were dropped from large armadas of melting icebergs. The ability of icebergs to carry large rocks into the ocean has been known for a long time, but the great size of Heinrich's armada was surprising. (Photo by Gary Comer)

Figure 4.6
Icebergs in Scoresby Sound. The open North Atlantic may have looked like this during Heinrich events. (Photo by Philip Conkling)

was that, as is the case in the basin today, the warmer interstadials were characterized by anoxic conditions, which killed off any worms that might have stirred the sediment, preserving the seasonal laminations. In contrast, during the cold stadials, waters sufficiently rich in oxygen must have ventilated this basin such that worms were able to inhabit the sediment and thereby stir it, thus destroying the laminations.

Kennett's breakthrough was soon matched by findings by other investigators documenting full sets of Dansgard–Oeschger events, as well as the Younger Dryas event, in sediments from other parts of the world. The key was to core sites where the sediment accumulated rapidly. One such site was in the Indian Ocean off Pakistan, where the record showed that the strength of the Asian monsoons jumped up and down in concert with the Greenland record. Cold stadials correlate with weak monsoons, warm interstadials with strong monsoons. This confirmed the dust and methane records from the Greenland ice cores, as described in the previous chapter. A marvelous oxygen isotope record in stalagmites from caves in China further verified this finding (see chapter 6). The correlation of records from different places suggested that the abrupt climate changes were not local events, but had propagated across major parts of the globe.

Other Paleoclimate Puzzles from Around the World

Complicating this situation was a discovery in the late 1980s by Hartmut Heinrich, a young German scientist. In the section of a deep-sea sediment core from the eastern North Atlantic representing the last period of glaciation, he documented a series of six layers rich in rock fragments, and nearly devoid of the foraminifera shells, which dominate the sediment between these layers. He concluded that the material in these layers must have been dropped from large armadas of melting icebergs. Glaciologist Doug MacAyeal postulated that these icebergs originated on the west side of the Atlantic when the lobe of Canada's Laurentide Ice Sheet (centered over Hudson Bay) catastrophically collapsed, sending a cascade of ice out through Hudson Strait into the northern Atlantic Ocean.

Subsequent studies of a series of cores stretching from the mouth of Hudson Strait to the site of Heinrich's core confirmed this seemingly outrageous hypothesis. Consistent with a Canadian source interpretation, the layers of ice-rafted debris thinned to the east, and isotopic measurements showed that rock fragments were likely derived from the ancient Archean terrain of northeast Canada.

It was puzzling, however, that although the impacts of these so-called Heinrich armadas show up at a number of the far-field sites, they had no distinctive correlatives in Greenland ice (or, for that matter, in the Santa Barbara basin). In contrast, in the Pakistan marine sediment and the Chinese cave stalagmite records, the Heinrich impacts are more pronounced than those associated with Dansgaard–Oeschger events. Furthermore, a record from Florida and two records from eastern Brazil reveal Heinrich impacts (and that by the Younger Dryas) but no hint of Dansgaard–Oeschger events.

So paleoclimatologists are in a situation somewhat akin to that faced by particle physicists. They have a zoo of fundamental units of matter and we have a zoo of abrupt events. Each group of scientists strives for a unified theory to account for all the inhabitants of its zoo. Paleoclimatologists divide the animals in their zoo into two distinct groups, the Dansgaard–Oeschgers (i.e., DOs) and the Heinrichs (i.e., Hs). But it is not clear into which group the Younger Dryas (i.e., YD) should be placed. In Florida and Brazil, where no DO signature is seen, the YD appears along with the Hs. Yet, in Greenland where no signature of the Hs is seen, the YD appears along with the DO events. This being the case, perhaps the YD should be placed in a category all its own. One might ask, why not use the presence or absence of an ice-rafted debris layer in North Atlantic cores as the basis for a decision? But again, the evidence is mixed. Although no YD debris layer is found in cores from the open Atlantic, a debris layer with about the right age is present in cores adjacent to the mouth of Hudson Strait.

A major clue regarding which animals belong with which in the zoo comes from the record of the Younger Dryas in Greenland. Based on isotopic measurements made on nitrogen and argon gas trapped in bubbles of the GISP2 ice core from Greenland Summit locale, Jeff Severinghaus, of the Scripps Institution of Oceanography, has demonstrated that the mean annual temperature in Greenland during the Younger Dryas cold snap was about 16°C colder than

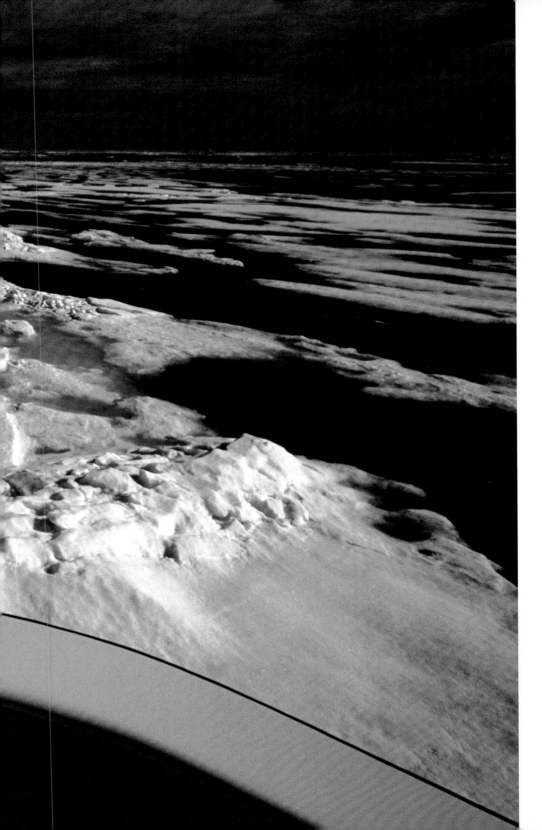

Figure 4.7
Extensive sea ice cover encountered on voyage to the northwest coast of Greenland, 2001. The seabird off the bow is a northern fulmar. (Photo by Gary Comer)

today's. In contrast, as discussed in chapter 3, based on the extent of lowering of the Younger Dryas snow line in Scoresby Sund, George Denton concludes that summer temperatures during the Younger Dryas were only about 4°C to 6°C colder than now.

As both of these estimates appear to be firm, in order to explain the large difference between them Denton suggested a seasonal hypothesis—that, during the Younger Dryas, winter temperatures in Greenland were 25°C to 30°C colder than now and summer temperatures only moderately cooler by 4°C to 6°C. Today, only Siberia experiences such a large summer-winter temperature difference. To make Greenland like Siberia, Denton concluded that, at least during the winter months, Greenland must have been entirely surrounded by sea ice. In this case, no heat could have escaped from the sea and the local sunlight would have been largely reflected back to space. Denton's sea-ice mechanism nicely explains, for example, Dorothy Peteet's Alaskan record. Further, the cold conditions would have enhanced not only the areal extent of winter snow cover, but also its duration, which would have delayed the onset of Asia's monsoons and in this way could account for their weakening.

Support for Abrupt Climate Change and a False Lead

U.C. Berkeley's John Chiang realized that the extensive sea ice cover in the northern Atlantic could provide an explanation for the tropical impacts of the Heinrich and Younger Dryas cold snaps. Based on a simulation carried out using a computer model, Chiang showed that the Northern Hemisphere cooling created by the extensive sea ice would push the thermal equator and its associated rain belt to the south. This shift allowed him to explain why normally dry eastern Brazil experienced high rainfall during the Heinrich events and the Younger Dryas. In this way Chiang explained the bursts of stalagmite growth and the increase in runoff from rivers in eastern Brazil.

Although a shutdown of deep-water formation combined with the formation of extensive winter sea ice appears to explain the very wide geographic extent of Heinrich event and Younger Dryas impacts, we still need an explanation as to what triggered these abrupt coolings and what terminated them in an equally abrupt manner. For the Heinrich events, the trigger has already been alluded to, namely the melting of a huge armada of icebergs. The reduction in salinity generated by their melting created a lid of low-density water akin to that which currently caps the water column in the Arctic Ocean. In this situation, winter cooling would not be able to create sufficiently dense water to sink to the abyss. Instead, in the absence of heat from the isolated deep water, the surface would freeze over. In addition, the warm water of the conveyor's upper limb no longer would be drawn in to replace that which sank to the abyss. And so the ocean would lock into a new conveyor "off-mode" of operation. But just as in the Arctic today, each summer sea ice would recede, and each succeeding winter it would reform.

Early on Broecker was enamored with the idea that the Younger Dryas was triggered by the catastrophic release of water stored in a lake that occupied the moat surrounding the retreating Laurentide ice sheet. He also considered the Younger Dryas to be a one-time event tagged onto the end of the last ice age. But, as with many of his initial thoughts on this subject, he eventually abandoned this idea. Its demise came about when the search using Gary Comer's Caravan airplane failed to reveal any geomorphic evidence along the route of the proposed flood. No channels, no boulder fields, were to be found. Further, the oxygen isotope record in China's Hulu Cave revealed that the sequence of events that occurred at the end of a previous time of deglaciation (known as Termination III) showed distinct similarities to the Bølling–Allerød-Younger Dryas oscillation. Hence a more likely candidate for a Younger Dryas trigger appears to be a seventh Heinrich ice armada.

Restarting the Ocean Conveyor

When Broecker first became interested in these abrupt changes, a German expert in ocean dynamics emailed him the following comment: "Wally, I can understand your claim that a jolt of fresh water initiated the abrupt coolings, but clearly you can't use similar jolts to bring them to an end." Of course, his point was well taken, for the transitions from cold back to warm were even sharper than the warm to cold transitions. Modelers soon supplied an answer. Using heat stored up in the deep sea allowed the fresh water lid in the North Atlantic off Greenland to be penetrated and thereby spontaneously restart the conveyor.

Although the jolts of fresh water may have been responsible for stopping the conveyor circulation at the times of the Heinrich armadas, it seems a stretch to call on such inputs to trigger each of the Dansgaard–Oeschger interstadials. Having come up with ways to bring to an end the cold Heinrich stadials (and also the Younger Dryas), the ever-inventive modelers found ways to create spontaneous stoppages as well as start-ups of the conveyor. Broecker was one of the first to propose a possible mechanism of this type. In order to explain why the surface waters in the Atlantic had one to two grams per liter more salt than their equivalents in the Pacific, he proposed that Earth's wind systems exported more water vapor from the Atlantic Ocean to the Pacific than they imported. The consequent buildup of salt in the Atlantic was balanced by export by the salty lower limb in conveyor waters. It was then only a small step to envision that if the export of water vapor were to get out of balance with the export of salt, the result would be stops and starts of conveyor circulation.

But this still does not explain the difference between DO and H events. Apparently, both involved stopping and starting of the conveyor. So what's the difference? Although the answer to this question remains a mystery, the difference probably lies in how the ocean operated when the conveyor was off. Stefan Rahmstorf, a modeler in Potsdam, Germany, envisions two different modes. During the H stadials, deep-water formation in the northern Atlantic came to a complete halt. During DO stadials, a second mode of deep-water formation kicked in. The location of deep-water production moved to the south 10 degrees and instead of sinking all the way to the bottom, it sank only about halfway.

But enough of this. The fact of the matter is that while we know that stops and starts of the ocean conveyor must have led to the alphabet zoo of abrupt DO, H, and YD climate events, much remains to be learned about the underlying physics.

Figure 4.8
The search for the flood channels to support the hypothesis of a northern outlet of the Lake Agassiz flood that could have triggered a shutdown of the ocean conveyor failed to locate supporting evidence. (Photo by Gary Comer)

Will the Ocean Conveyor Stop?

One other aspect of this subject merits comment. Is there any chance that the ongoing buildup of CO_2 and other greenhouse gases will result in a conveyor shutdown? At one time, this seemed likely, but, as did many of Broecker's early thoughts, this one has evaporated. The idea was that perhaps the increase in rainfall and river runoff created by the warming would freshen the northern Atlantic to the point where the conveyor would halt. However, simulations in models suggest that only in the case of a very large increase in CO_2 (fourfold) would this happen. Further, the shutdown would not be abrupt, but rather would stretch over more than 100 years. And even if the conveyor were to shut down, the consequences would be far less severe for it would be far too warm for sea ice to form. Without sea ice, there would be no large amplification of the impact. So, in Broecker's mind, the concern regarding a future conveyor shutdown should be moved well down the list of possible consequences. But, because the ocean models are not perfect, we probably cannot eliminate a shutdown from the list of consequences. And, as explained in chapter 5, wobbles in the conveyor may make a big difference to the people who live along its path.

A WOBBLY NORTH ATLANTIC CONVEYOR?

The Little Ice Age

In the 1930s Francois Matthes, working for the United States Geological Survey, studied very young moraines deposited by small glaciers in the Sierra Nevada of California, and during an interview with a journalist introduced the term *Little Ice Age* to describe the time interval during which these moraines were deposited. The term has endured ever since, although its usage has varied. The modern application of the term generally applies to a relatively cold interval over the last few centuries, marked by expansion of mountain glaciers. Hubert Lamb, the great climate historian, defined the Little Ice Age as the period when "not only in Europe but in most parts of the world the extent of snow and ice on land and sea seems to have attained a maximum as great as, or in most cases, greater than, at any time since the last major ice age." The period that Lamb referred to was approximately 1550 to 1850.

George Denton's first encounter with the Little Ice Age occurred in the early 1960s in the northern reaches of the St. Elias Mountains of southwestern Yukon Territory. These spectacular mountains comprise the highest peaks in Canada, including Mt. Logan at 5,959 meters, and harbor the largest ice field in North America. Long outlet glaciers drain this ice field in star-like fashion.

On the south, many of these outlets calve into the Gulf of Alaska. But in the north, these outlet glaciers end on land in glacially carved valleys. One of these outlets is the beautiful Kaskawulsh Glacier, which terminates at the head of the valley occupied by the braided outwash stream of Slims River.

Denton landed in a bush plane with his field party on the Slims outwash plain near the glacier terminus, where they established a field camp with Dahl sheep, grizzly bears, moose, and caribou to keep them company. They mapped the recent moraines deposited a kilometer of so in front of the terminus and found spruce logs embedded in the outermost fresh-appearing ridge, including the remains of one tree that was still rooted but had been tipped over and sheared by advancing ice. Radiocarbon dates of these trees showed that the glacier advance that created these moraines had occurred within the last several centuries.

In the last few years, Alberto Reyes and Brian Luckman revisited the moraines in front of Kaskawulsh Glacier and, from a comparison with a regional tree-ring data base, placed the glacier kill dates of these trees in the mid-eighteenth century, thus obtaining a more precise age for the outermost moraine of the Little Ice Age.

But to return to their earlier study of the 1960s, Denton and his field group examined the valley floor just outboard of the young Kaskawulsh moraines in areas not covered by Slims outwash. Here they found a thick layer of wind-borne *loess* deposited during the retreat of Kaskawulsh Glacier from Slims valley at the end of the last ice age, and another layer from the recent glacier resurgence in the second half of the ongoing interglaciation. Grass buried at the base of one of the exposures of loess yielded a radiocarbon date of 11,200 calendar years ago, which was significant because the section exposed only windblown loess, with the conspicuous absence of any glacial deposits that would indicate an overriding of the site by Kaskawulsh Glacier. The fact that this loess deposit nestled right up next to the moraine with the radiocarbon-dated wood meant that, within the last few centuries, Kaskawulsh Glacier attained its maximum extension of the last 11,200 years

Thus, in the case of Kaskawulsh Glacier, the Little Ice Age advance of the last few centuries represented the greatest glacial advance since the close of the last ice age, just as Lamb envisioned. This situation immediately raised the interesting question of what the Little Ice Age—the coldest time of this

Figure 5.1
Long outlet glaciers drain Alaska's ice fields, many flowing into the Gulf of Alaska. (Photo by Gary Comer)

interglaciation—is trying to tell us about the progression of climate over the ongoing interglaciation.

The Little Ice Age in Europe

Denton's next encounter with the Little Ice Age occurred in the late 1960s, when, papers and maps in hand, he visited the classic glaciers of the Swiss Alps and those of southwest Norway. Here he saw the evidence of fluctuations of mountain glaciers during the Little Ice Age amply documented by paintings, lithographs, written descriptions, and tax records of real estate in the path of advancing ice fronts, as well as by actual measurements of the changing positions of glacier termini.

Such records show that in Switzerland the Little Ice Age had two prominent maxima, one in the 1600s and the second, sometimes referred to as the *Hochstand*, in the mid-nineteenth century. Widespread retreat from the Hochstand began in the 1850s to the 1860s, bringing the Little Ice Age in the Swiss Alps to a close. Glacier recession continued into the 1960s, with the exception of a "still-stand" or minor readvance that occurred in the 1920s. By the 1960s a second still-stand was underway for glaciers in the European Alps and as far afield as the western United States. It was this second advance that led to news stories about the onset of a new ice age.

In Norway, the structure of the Little Ice Age was a bit different from its signature farther south in the European Alps. For example, in southwestern Norway the maximum extent of Nigardsbreen, an outlet glacier of the Jostedals ice cap, occurred earlier, in 1748, very similar in timing to the maximum extension of the Kaskawulsh Glacier in southwestern Yukon Territory.

As evidence began to accumulate from mountain glaciers elsewhere in the Northern Hemisphere, particularly in western North America, the notion that the advances of the Little Ice Age represented the maximum cold pulse since the end of the last glaciation became stronger and stronger. Tree-ring chronologies of glacier fluctuations in southern Alaska, for example, first carried out by Donald and Elizabeth Lawrence more than a half century ago, and more recently by Greg Wiles and his colleagues, indicate that the maximum of the Little Ice

Figure 5.2
Grizzly bears fishing in glacial melt water streams enliven geological field research expeditions. (Photo by Gary Comer)

Age in this sector of the planet was achieved in the middle of the eighteenth century. In the Canadian Rockies, Brian Luckman concluded that the overall maximum glacier extent probably dated to 1830–1850 and was followed by recession that brought the Little Ice Age to a close. This situation turned out to be similar to that farther south, in the U.S. Rockies, where the maximum of the Little Ice Age also seems to have occurred at about 1850. Likewise, both in Iceland and on Baffin Island in the Canadian Arctic, the Little Ice Age advances also were the most extensive since the end of the last ice age.

Recognition of the importance of the Little Ice Age prompts an obvious question. Was the Little Ice Age an isolated event, or was it preceded by other similar but less extensive pulses during the long ongoing interval of interglacial climate? Such a question is really aimed at deciphering the underlying cause of the Little Ice Age and any of its predecessors and also begs the related question of why the Little Ice Age was the maximum glacier pulse of this interglaciation, at least in the Northern Hemisphere. The implication is of cooling interglacial climate. If so, why? And would the answers to these questions give us clues to the long-standing ultimate question of how ice ages are initiated under natural conditions?

The Little Ice Age in Alaska and the Alps

To attack this problem, Wibjorn Karlen and George Denton returned in the early 1970s to southwestern Yukon Territory and adjacent Alaska along the northern slope of the St. Elias and Wrangell Mountains. They also carried out field work in Lapland near Sweden's highest peak, Kebnekaise, and in Sarek National Park, which both featured numerous small mountain glaciers.

In Alaska, Yukon Territory, and Lapland they specifically searched for sets of glacier moraines where large old ridges formed topographic barriers in front of subsequent ice advances, which either confined these advances or else deflected them away from moraines deposited during older advances. These special geomorphologic situations preserved the old moraines from overriding and destruction during the Little Ice Age. As a result, in those places they could distinguish earlier advances even though they occurred at times when climate was not quite as cold as during the Little Ice Age.

The results of this field work in southern Alaska and Yukon Territory, as well as in Lapland, showed that the Little Ice Age was not an isolated occurrence, but was the last and most extensive in a succession of similar episodes of glacier expansion and retreat that punctuated this interglaciation. Karlen and Denton were able to place a rough chronology on this succession of "little ice ages" using radiocarbon dating of organic matter such as trees buried beneath moraines. However, this procedure did not yield the precision needed for comparing glacial chronologies with other climate proxies. For that precision, they needed to look to Switzerland once again.

For nearly a quarter of a century Hanspeter Holzhauser has monitored retreating glaciers in the Swiss Alps. He has concentrated his studies on the Great Aletsch, Gorner, and Lower Grindelwald glaciers, all of which have historical records that extend well back beyond the Little Ice Age. Because these glaciers projected below tree line during at least the last several thousand years, changes in the three termini can be reconstructed in remarkable detail from dendrochronological analyses. As the termini of these glaciers underwent overall retreat during the twentieth century, Holzhauser located more and more remnants of old fossil trees that were emerging on the deglaciated terrain. Such trees had previously been crushed when, in past centuries, these same glaciers had undergone pulses of advance, but the tree remnants had survived until recent recession uncovered them again. Holzhauser could determine the year in which each tree had been killed by tying the tree rings of these remnants into a master regional tree-ring chronology. The product was an astonishingly accurate chronology of glacier fluctuations in the Swiss Alps over the past three and a half millennia.

Holzhauser's results showed that the Little Ice Age was more complicated than had been previously recognized. An early phase of the Little Ice Age began in the fourteenth century and featured a major advance that culminated about 1369 for the Great Aletsch, 1385 for the Gorner, and 1338 for the Lower Grindelwald glaciers. These results also suggested that the late phase of the Little Ice Age began with a marked climate change about 1565 that led to extensive glacier expansion, with persistent advanced ice positions for almost three centuries. As noted above, the greatest extent of this interval was achieved during the early seventeenth century and again during the middle of the nineteenth. In summary, Holzhauser's results, combined with existing historical records, show that the Little Ice Age in the Swiss Alps lasted from 1300 until 1855–1865, with glacial maxima in the late fourteenth century in the early phase, followed in the late phase by maxima in the first half of the seventeenth century and then again in the middle of the nineteenth. In the late phase, ice margins stayed very close to their maximum extents for the entire time between the two peaks of ice extent.

The Little Ice Age in Western North America

Similar studies to those of Hanspeter Holzhauser have recently been completed in the cordillera of western North America in several localities where glacier termini extended below tree line during the ongoing interglaciation. One is in the coastal ranges of southeastern Alaska. Here Greg Wiles and his colleagues recognized both the early and the late phases of the Little Ice Age. The former took place between 1180 and 1300. Ice expansion during the late phase culminated in the early seventeenth century, with subsequent maxima during the eighteenth and nineteenth centuries. The greatest advance occurred during the middle of the eighteenth century, a similar situation to that discussed earlier for the Kaskawulsh Glacier on the north flank of these same coastal ranges. In any case the Alaska coastal glaciers remained close to their mid-eighteenth-century position until as late as 1890, after which widespread recession set in.

Wiles also recognized earlier events in southeastern Alaska that were similar to those described by Hanspeter Holzhauser in the Swiss Alps. An equivalent to the Medieval Warm Period culminated between 900 and 1100. Furthermore there was a major glacier advance during the time of the Dark Ages in Europe. In fact, Alberto Reyes and his colleagues assembled data that show that an equivalent of the Swiss "Dark Ages" advance was widespread throughout the mountains of western Canada.

Applying tree-ring techniques, Brian Luckman and his colleagues also discovered both an early and a late phase of the Little Ice Age in the Canadian Rockies. The prolonged early phase encompassed the times of both the

Figure 5.3
Alaska's coastal glaciers remained close to their eighteenth-century position until the late nineteenth century when widespread recession set in, including in Geographic Bay shown here. (Photo by Gary Comer)

twelfth- and the fourteenth-century advances in the Swiss Alps. The late phase featured two widespread pulses of glacier expansion, one at about 1700–1735 and the other between 1830 and 1859. Luckman estimated that the overall regional ice cover in the Canadian Rockies probably dated to the mid-nineteenth century and was followed by recession that brought the Little Ice Age to an end. Luckman also discovered evidence in the Canadian Rockies at Peyto Glacier for a "Dark Ages" advance, as well as for an expansion about 2,800 years ago during what is called by Holzhauser the Iron Age Cold Epoch in the Swiss Alps.

The chronological results of Wiles and Luckman illustrate that the Little Ice Age signature in western North America is similar in most ways to that in the Swiss Alps. Moreover, older Holocene oscillations appear to replicate in both Europe and North America as well, at least within the margin of dating errors. These similarities suggest a widespread climate signal in the middle latitudes of a significant swath of the Northern Hemisphere. A further implication of the overall record is that the Little Ice Age was not an isolated occurrence, but rather was the last in a succession of similar events registered by deposits of mountain glaciers in the Northern Hemisphere.

The Little Ice Age in Greenland

This was the state of scientific discussion when Gary Comer invited Denton and his colleagues to join him on a series of cruises along the Greenland coast from 2003 to 2006 on his expedition vessel, the *M/V Turmoil*. The two cruises to Scoresby Sund were of particular interest, because this region of Greenland lies alongside the cold East Greenland Current, the primary oceanographic feature that determined the extent of sea ice in the northern North Atlantic during the Little Ice Age. The 2005 cruise to South Greenland was equally fascinating, because the region was riddled with the ruins of Norse colonies set up in the late 900s and occupied until sometime in the fifteenth century.

What they found in Scoresby Sund was very similar to what Denton had seen at the Kaskawulsh Glacier four decades earlier. Fresh-looking and conspicuous moraines occur 1–2 kilometers beyond the snouts of nearly all ice lobes and are commonly associated with fresh "trimlines." Such moraines and trimlines elsewhere in Greenland are dated by historical records to the middle or late nineteenth century, although a few may have been deposited in the early nineteenth or middle twentieth century. They mark what is referred to as the *historical ice limit* and almost certainly date to the Little Ice Age.

Just as at Kaskawulsh Glacier, the fresh-looking moraines in East Greenland represent the maximum glacier extent since the end of the last ice age. Brenda Hall of the field party determined this important fact by mapping shell-laden marine sediments resting in valleys that had become ice free at the end of the last ice age. These shells are in their current locations above present-day sea level because of land upheaval, or glacial rebound, that occurred after the removal of the ice load of the last glaciation.

The ages of these shells were determined by the radiocarbon dating method, and the results show that shells date to more than 10,000 years ago in marine sediments that extend the length of the valleys right up to the historical moraines. These shells and the enclosing marine sediments exhibit no signs of being overridden by glacier ice. The implication of this discovery was that East Greenland glaciers had never been more extensive in the last 10,000 years than they were when they reached their historical limit during the Little Ice Age.

In southern Greenland, a similar historical limit occurs within several kilometers of the ice-sheet margin. Moreover, there is evidence that the southern sector of the Greenland Ice Sheet seems to have been expanding through the second half of this interglaciation. Such expansion of the ice sheet is consistent with the results shown in a pollen diagram from East Greenland, which indicates a long-term cooling trend after an early interglacial optimum, which culminated in the Little Ice Age. Such a trend is also in accord with the frequency of melt layers in the GISP2 ice core in central Greenland, which has been steadily decreasing through much of the interglaciation. Moreover, the inversion of the borehole thermal data from central Greenland shows general cooling since the middle of the interglaciation through to the Little Ice Age.

All of these climate indicators from Greenland are in agreement with a number of paleotemperature proxies that, taken together, indicate that Arctic lands have cooled after an early interglacial warm interval, the timing of which varied from place to place. For example, the northern tree line in Russia has been retreating. Moreover, after an early interglacial optimum, the elevation of the Scots pine tree line in the Scandinavian mountains has steadily lowered,

Figure 5.4
Fresh-looking and conspicuous moraines occur beyond the snouts of nearly all ice lobes and are commonly associated with fresh trimlines in Greenland, including here at the upper end of the Qinguadalen valley. (Photo by Gary Comer)

signifying a cooling trend in summer temperatures that culminated in the Little Ice Age.

From its base camp in Scoresby Sund on the *M/V Turmoil* in the summer of 2003, the party took many reconnaissance flights throughout East Greenland with Gary Comer in his Caravan floatplane. Telford Allen, a long-time pilot for Gary Comer and neighbor of Denton's from Bangor, Maine, was at the controls. One such flight passed over the mountains and plateaus of Liverpool Land, located alongside Scoresby Sund near the edge of the Greenland Sea. At the last glacial maximum, Liverpool Land supported a large ice cap that coalesced with inland ice. Now, however, all that was left was a series of small remnant ice caps on the inland plateau, along with small mountain glaciers in the cirques in the nearby mountains.

On this flight across the Liverpool plateau the party spotted a spectacular example of collapsing ice caps. It was late in August near the end of summer ablation season. No snow remained on the surface of these small ice caps, so that they had no annual nourishment. They appeared to be dying. Dirty ice bands cropped out on the surface of each ice cap in a pattern that looked like concentric tree rings around the center of each ice cap as annual summer layers became exposed by ablation of the surface of the ice cap to form a circular pattern when viewed from above.

The thinning and dying ice caps revealed newly deglaciated terrain alongside their retreating margins. Thus the situation was perfect for examining the ground near the edges of these ice caps for the emergence of any organic material that may have remained preserved by ice during the Little Ice Age in order to discover when the ice caps had last sagged back to their current configurations. The chances for preservation were high, since the ice caps were thin and almost non-erosive, even when they expanded during the Little Ice Age.

Led by Tom Lowell, one of the field parties investigated the terrain alongside the thinning ice caps of Liverpool Land that had been spotted from the air. What they found was spectacular. They first noted that the typical sharp trimline and fresh moraine of the historical limit of the Little Ice Age occurred a few hundred meters outboard of the present-day ice margins, as well as high on the sides of *nunataks* (hills completely surrounded by glacial ice) that project through the ice cap. Exposure dating of rocks on the moraine indicated that they indeed dated to the Little Ice Age. Furthermore, exposure dates on bedrock

Figure 5.5
Telford Allen, a long-time pilot for Gary Comer and a neighbor of George Denton's in Maine, was the floatplane and helicopter pilot supporting the field research expeditions in East Greenland. (Photo by Philip Conkling)

Figure 5.6
On the Liverpool plateau, no snow remained on the surface where all of the glaciers had completely melted away. In this image, a hilltop of mudstone is wasting away periglacially, leaving concentric circles of dirty bands that look like tree rings when viewed from above. (Photo by Philip Conkling)

Figure 5.7
Geologist Tom Lowell led one of the field parties to collect exposure dates on moraines and isolated boulders around the periphery of Scoresby Sound, including on the lower part of the Gurrenholm Dal pictured here. (Photo by Paul Migro)

just outboard of the fresh moraines showed that the Little Ice Age moraines represented the maximum extent of these ice caps since the close of the last ice age.

But the most striking find was the widespread preservation of old organic remains on the newly deglaciated terrain. These remains either were tucked into cavities in the bedrock or were preserved beneath rocks dropped during the passage of the ice caps over the organic remains. In a way these organic deposits could be viewed as a forest that was buried and then revealed again by the ice caps. But rather than large trees, the forest consisted of the very small Arctic dwarf birches and shrubs, along with many different plant species. The radiocarbon dates of this emerging organic material showed that between 800 and 1000 the Liverpool ice caps were at least as small as, or even less extensive than, they are now. These dates encompass the time when Erik the Red explored the Greenland coast and then established Norse colonies in south Greenland in the late 900s, which is also within the time that Hanspeter Holzhauser tagged as the Middle Ages Warm Interval in his reconstruction of glacier variations in the Swiss Alps. Although these results are consistent with glacier thinning and recession in Liverpool Land during the Middle Ages, we are quick to note that we cannot yet determine whether temperatures then were warmer or cooler than now. The reason is that the thinning ice caps have not yet come into balance with the present-day climate. Therefore, we do not know how much retreat is yet to occur. In fact, some of the ice caps seem destined to disappear entirely. And as yet we only have a minimum value as to how much recession occurred during the Middle Ages.

Greenland Norse and Sea Ice

Figure 5.8 depicts sea-ice configuration near Iceland through the last millennium, taken from the reconstructions of Lamb, Koch, and Ogilvie and Grove, based on historical records of sea ice from various Norse sources. There is a general consensus from circumstantial evidence that sea ice was sparse along the medieval sailing routes followed by the Norse across the North Atlantic

Figure 5.8
Map of North Atlantic region showing the maximum extent of spring sea ice during the latter part of the Little Ice Age (light blue regions bound by stippled lines), and medieval Norse sailing routes between Norway, Iceland, Greenland, and North America (red lines). Red shading shows general locations of Viking settlements in Greenland during medieval time. Bar graph at bottom depicts sea-ice extent near Iceland over the last ~1,340 years. Figure adapted from G.H. Denton and W. S. Broecker, Wobbly ocean conveyor circulation during the Holocene? *Quaternary Science Reviews* 27 (2008): 1939–1950 and references therein.

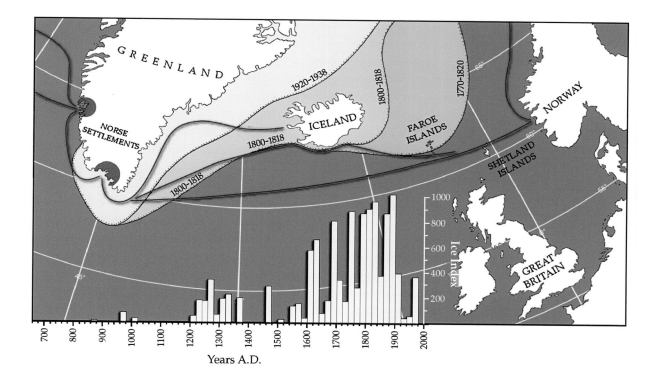

between 800 and 1170 during their settlement of the Faroe Islands, Iceland, and Greenland. Koch estimated that sea ice returned to the vicinity of Iceland in the thirteenth and fourteenth centuries.

The first historical reference to a changing climate in Greenland describing the problem of sea ice is in a Norwegian source, *The King's Mirror*, written in 1260. As cited by Seaver, the book indicates the climate in Greenland had already begun to decline by the middle of the thirteenth century. "There is more ice to the northeast and north of the land (on the east coast of Greenland), ... consequently whoever wishes to make land should sail around it to the southwest and west till he has come past all those places where ice may be looked for and approach the land on that side." Sea ice may have made voyages to

Labrador (which the Norse called Markland) for timber and iron blooms more and more hazardous.

Isotope studies reported by Lynnerup show shifts over time in the O_{18} proportions from human tooth enamel taken from the remains in graveyards in the Norse region known as the Eastern Settlement (actually in southern Greenland), which reflects the temperature at the time the enamel is first formed in infancy. These studies suggest that the temperature in the Eastern Settlement had begun to decline in the twelfth century and continued downward for several hundred years.

Ogilvie's recent examination of documentary evidence from Icelandic sources concluded that the climate there was deteriorating by the late twelfth century and was particularly severe in 1306 and 1319–1321. Anne Jennings and Nancy Weiner reported on sediment-core data reflecting sea ice in Nansen Fjord in southeast Greenland that demonstrate changes in polar currents and an increase in sea ice formation during the fourteenth century, peaking in 1370. In addition, temperature-sensitive deuterium isotope records published by Lisa Barlow show yearly low temperatures between 1308–1319 and 1324–1329, mostly during the winter. However, the period between 1342 and 1362 is especially notable, according to Barlow, because it was marked primarily by colder summer temperatures. It is important that these colder summer temperatures coincide closely with the period of time historians and archeologists believe the Norse abandoned the Western Settlement.

During that interval, numerous polar bear skins are recorded in church inventories, indicating that bears were present on near-shore ice. Milder climate prevailed between 1412 and 1470, but severe conditions with many years of heavy ice had set in by the last decade of the fifteenth century. Nevertheless, the historical, archeological, and scientific records indicate that the main part of the Greenland Norse in the Eastern Settlement survived long periods of cold during deteriorating climate conditions throughout the 1300s before this larger settlement vanished sometime after 1450.

In the North Atlantic near Iceland, sea ice was common during much of the late phase of the Little Ice Age between about 1600 and the middle to late part of the nineteenth century, after which sea ice became much less common, with one conspicuous exception: a spike of sea ice near Iceland between 1965 and 1971.

What Caused the Little Ice Age?

There has been much scientific speculation in the literature as to what would have caused the Little Ice Age and its predecessors. Proposals include either random fluctuations of climate or else volcanic activity. But perhaps the most common theme is that such glacier fluctuations have been influenced by variations in solar output. Such inferences come largely from what appear to be coeval changes in glacier extent with changes in the production rates of both ^{10}Be and ^{14}C, which can in theory be related to solar output.

Edouard Bard, for example, constructed a solar irradiation curve from such variations in production rates, concluding that changes in solar irradiance were compatible with cooling during the Little Ice Age. But there are problems with this model. One is the paucity of climate records detailed enough to compare with the inferred changes in solar irradiance. But most important, a comprehensive mechanism whereby small changes in solar irradiance can be transferred into such a large, widespread climate signal has yet to emerge.

The indication of a Medieval Warmth/Little Ice Age oscillation in Greenland raises the question of the cause of the millennial-scale climate oscillations that are superimposed on the overall climate deterioration of this interglaciation. After looking out over the Greenland Sea from Liverpool Land and from the mouth of Scoresby Sund, the team began to wonder if these millennial-scale climate oscillations were driven by changes the North Atlantic currents that lay at the northern end of the great ocean-wide conveyor circulation.

The team first recognized that the cold East Greenland Current flowed southward along the east coast of Greenland, then westward around Greenland's south coast, and finally northward into the Labrador Sea. An important branch of this current, dubbed the East Icelandic Current, peels off of the East Greenland Current directly offshore of Liverpool Land and Scoresby Sund and flows southeastward toward northern and eastern Iceland. The outer edge of the East Greenland and the East Icelandic currents defines today's polar front in the North Atlantic. The North Icelandic Irminger Current, a branch of the relatively warm North Atlantic Drift, flows northward up the west coast of Iceland and then curls eastward around the north coast. The Icelandic Current, an offshoot of the North Atlantic Drift, now flows eastward along the southern coast of Iceland.

Recent results from a high-resolution record in a marine-sediment core taken off northern Iceland strengthen the conclusions drawn from the historical records of sea ice. The core has a high rate of sedimentation. The chronology derives from the occurrence of volcanic ashes in the core of known age. The core contains the remains of a unique organic compound produced by diatoms living in sea ice. This biomarker shows a sea-ice history that is similar to that derived from historical sea-ice records.

Little sea ice is indicated through the Middle Ages until 1300, when the frequency began to rise, with particularly severe ice about 1309, 1331, and 1364. There was another sea-ice pulse in the second half of the fifteenth century, but ice was relatively rare from then until the end of the sixteenth century. High values of sea ice returned to the waters off of northern Iceland during the late phase of the Little Ice Age. Alkenone paleothermometry from the same core produced sea-surface temperatures back through the past two thousand years. The results basically produce the same record as obtained previously from the sea ice biomarker, but they extend farther back in time. Warm pulses occurred in the Roman Warm Period and the Medieval Warm period, separated by colder temperature during the Dark Ages and the Little Ice Age.

Taken together, the glacier records from Switzerland, the glacier signature from the Canadian Rockies, the glacier records for coastal southern Alaska, the sea-ice indices from near Iceland, and the sea-surface temperatures of the waters off of northern Iceland show pronounced similarities. The major oscillations in climate recognized by Hanspeter Holzhauser in the Swiss Alps appear in the Icelandic sea-ice and sea-surface record, as well as in the glacial record from the western cordillera of North America.

Figure 5.9
Iceberg flotillas departing from the mouth of Scoresby Sound enter the East Greenland Current, which flows south, or join another branch, the East Icelandic Current, which splits directly off Scoresby Sound and flows toward Iceland. The outer edges of these currents mark the present location of the polar front in the North Atlantic. (Photo by Philip Walsh)

A Common Climate Signal

Thus there appears to be a common climate signal across a considerable swath of the Northern Hemisphere. Of course, such a common signal could simply be a response of the sea ice and mountain glacier systems to common climate forcing from the sun or from volcanic eruptions. But it could also reflect an internal element of the climate system, most notably the strength of the conveyor

Figure 5.10
Comparison between various instrumental and proxy temperature records from the North Atlantic region over the last ~120 years. Top panel shows detrended, normalized sea-surface temperatures (SST) integrated over the North Atlantic region. Blue indicates periods of cooler SSTs and red represents periods of relatively warmer SSTs. Middle panel shows instrumental air-temperature data from various Greenland sites. Bottom panel depicts the number of Swiss glaciers advancing (blue) retreating (red), or remaining stationary (yellow). Black vertical lines mark coeval transitions from cooler- to warmer-than-average conditions in and around the North Atlantic. (Figure adapted from G. H. Denton. and W. S. Broecker, Wobbly ocean conveyor circulation during the Holocene? *Quaternary Science Reviews* 27 [2008]: 1939–1950 and references therein, and A. J. Long, Back to the future: Greenland's contribution to sea-level change, *GSA Today* 19 [2009]: 4–10.)

circulation in the northern North Atlantic Ocean. In other words, the question the team began to contemplate is whether wobbles in the strength and position of oceanographic currents of the North Atlantic conveyor lie at the root of the millennial-scale fluctuations of Holocene climate.

About 15 years ago, an interesting discovery was made concerning the North Atlantic conveyor by Michael Schlesinger and Navin Ramankutty, who postulated that the ocean-atmosphere system had an internal oscillation with a period of 65–75 years superimposed on the well-known overall warming since the middle of the nineteenth century. Sea-surface temperature varied by about 0.4°C during a complete oscillation. It was further postulated that the core of this oscillation resided in the Atlantic Ocean north of the equator, and that the oscillation was linked to variability in the strength of thermohaline circulation. Richard Kerr, a writer for *Science,* dubbed this phenomenon the *Atlantic Multidecadal Oscillation* and referred to the North Atlantic as a wobbly ocean.

With further investigation, it was shown that warm phases of the Atlantic Multidecadal Oscillation occurred at 1860–1880, 1940–1960, and 1990–present. Cold phases occurred at 1905–1925 and 1970–1990. A physical argument was set forth that these variations in North Atlantic sea-surface temperatures, caused by variations in the strength of thermohaline circulation, affected summer temperatures at least in North America and Europe. It was even suggested that the imprint of the Atlantic Multidecadal Oscillation is widespread across the Northern Hemisphere.

The rhythm of the Atlantic Multidecadal Oscillation immediately resonated with the team. It is a dead ringer for glacier behavior in the Swiss Alps. For many years glaciologists have measured the fluctuations of the termini of 38 glaciers in the Swiss Alps. Since the middle of the nineteenth century, these glaciers have undergone considerable overall recession. But superimposed on this trend have been curious short-lived periods of expansion, separated by particularly pronounced pulses of recession. One of these periods of expansion occurred during the 1960s and 1970s, a time when, as we noted previously, sea ice reappeared near Iceland. These events led to speculation in some media articles more than four decades ago about the possible onset of a new ice age.

However, a glance at figure 5.10 shows a remarkable match between the fluctuations of glaciers in the Swiss Alps and oscillations of sea-surface records assigned to variations in the strength of the conveyor. The two cold phases of the

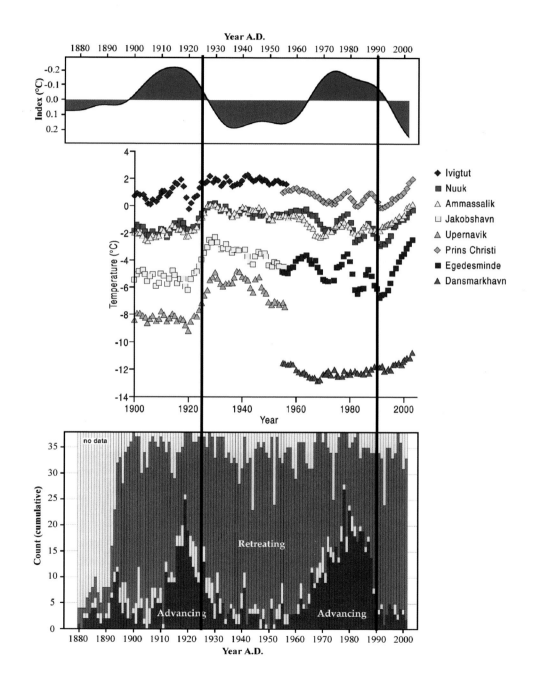

Atlantic Multidecadal Oscillation in the 20th century at 1905–1925 and 1965–1990 correspond with the two intervals of glacier expansion in the Swiss Alps. The North Atlantic warm phases at 1930–1960 and after 1990 match times of Swiss glacier recession. Thus it appears that, if the North Atlantic wobbles indeed affect summertime temperature changes in Europe, then they may also have been responsible for these oscillations of Swiss mountain glaciers. In addition to its effect on Swiss glaciers, the wobbly North Atlantic conveyor also seems to have had a pronounced influence on atmospheric temperature in Greenland. Presumably these temperature changes affected glaciers in Greenland just as they did in the Swiss Alps.

Such associations of North Atlantic wobbles over the past century with glacier and sea-ice pulses led to explorations of the possibility that the millennial-scale oscillations of glaciers throughout the course of the ongoing interglaciation were, likewise, associated with, and perhaps driven by, concurrent changes in the strength of the conveyor in the North Atlantic. In this regard, the sea-ice record in the North Atlantic strongly suggests a conveyor response for the Medieval Warm Period–Little Ice Age–twentieth century oscillation. It is tempting to speculate that the same relations held for the earlier Holocene climate oscillations recorded by the glacier fluctuations.

The sea-surface temperature data from north of Iceland certainly suggest that this is the case for at least the last 2,000 years, but a clear demonstration of such an association awaits the emergence of a sea-ice record for the entire interglaciation. In the meantime, the long-term cooling of the Northern Hemisphere evident from paleoclimate records may well reflect weakening of the North Atlantic conveyor, ultimately driven by declining Milankovitch summer insolation. Further, wobbles in conveyor strength superimposed on this decline may have been responsible for the millennial-scale oscillations of climate, including the Little Ice Age. Such oscillations could either be strictly internal or they could be paced externally; the important point is that the conveyor, which controls the distribution of sea ice around Greenland and Iceland, provides a comprehensive mechanism for understanding abrupt climate changes.

GREENLAND'S CLIMATE SIGNAL ACROSS THE GLOBE

With the discoveries of abrupt climate changes revealed in the Greenland ice cores and the evidence of big ice sheet collapses revealed in Heinrich's ice-rafted debris, Wallace Broecker began to wrestle with the question of how much of the rest of Earth might also have experienced these events. In particular, he began to think about the experiences of his early career in arid parts of the western United States.

Rainfall Shifts in Closed Lake Basins

Because Broecker grew up poor in the mid-west's flatlands and majored in math and physics, the subject of geology had never captured his attention. His first encounter with a geologist came during his second summer at Lamont. Professor Kulp found himself too busy to fulfill an obligation to present a lecture on the radiocarbon age-determination method to a group of archeologists who were to meet at the Southwest Museum in Los Angeles. At the last minute, he asked Broecker to go in his place. So, with sweaty palms, he gave his first scientific talk.

What Broecker remembers is not the lecture itself but what happened immediately afterward. A middle-aged man of the west with a potbelly and cowboy boots strolled up the aisle and faced him across the podium. "Young man, I can see that you know a lot about math and physics but clearly you don't know a gol darn thing about the Earth. Come with me for three weeks and I'll change your life." Quite a mouthful, but he was right.

After a bit of discussion about who would pay for the trip and a call to his wife, Broecker signed on and the next day he and Phil Orr were in Orr's aging station wagon on their way to Pyramid Lake, a jewel in the Nevada desert to the northeast of Reno. As is the case for Great Salt Lake, located on the eastern side of what is known as the Great Basin, Pyramid Lake has no outlet. As a result, all the water it receives must be eventually lost by evaporation. Consequently, these so-called closed-basin lakes fluctuate in size. If more rain falls into their catchment basins, they expand until the loss by evaporation once again balances the gain from runoff. With less rain, they contract in size.

Orr explained to Broecker that during the last glacial period Pyramid Lake and its sister remnant lakes in the area were once joined together to form a single huge lake. This greatly expanded glacial predecessor of Pyramid Lake carries the name Lahontan. Great Salt Lake's glacial big brother carries the name Bonneville. In both cases, the shoreline terraces that form bathtub-like rings around the present day remnant lakes provide evidence of the former greatly expanded lake.

In order to introduce Broecker to these paleoshorelines, Phil Orr secured a boat and took him to Pyramid Lake's Anaho Island (figure 6.1). They climbed to its crest, which is ringed by the highest of Lahontan's shorelines. There for the first time he encountered *tufa*, an algal deposit of $CaCO_3$ formed on rocky shorelines exposed to sunlight. He eagerly broke off a piece—his first geological specimen! Later he determined its radiocarbon age.

Over the next few summers, Phil and Wally collected samples from a number of locales in both Lake Lahontan and Bonneville basins. They obtained a host of ^{14}C ages. Later on in the 1960s, Archie Kaufman and then still later in the 1980s, Jo Lin did PhD theses under Broecker's direction investigating the possible use of $^{230}Th/^{234}U$ ratios as an alternative means of dating $CaCO_3$ material from the shorelines of these lakes. But it wasn't until 2005, half a

Figure 6.1
Photograph of Pyramid Lake's
Anaho Island.

century after his first encounter, that Broecker realized that the information contained in these lake records had far greater value than he had imagined.

Held's Hypothesis

Ultimately Broecker's eyes were opened while he listened to a lecture by Isaac Held, an atmospheric dynamics expert. Based on thermodynamic considerations and backed by careful modeling, Held made the case that as the world warmed as a result of the continuing buildup of fossil fuel CO_2, rainfall would be ever more strongly focused in the tropics and, as a consequence, the extra tropical dry lands would become ever more arid. As 40 percent of the world's grain is grown on irrigated land and as 1.5 billion of the world's poorest people reside on dry land, it was clear that Held's prediction, if correct, would have dire consequences. Hence, its validity requires careful evaluation.

It occurred to Broecker that even though no adequate warm analog exists in paleoclimatic records with which to test Held's hypothesis, there is a cold one, that is, glacial times. To the extent that Lahontan and Bonneville typify the extra-tropical dry lands that Held predicted would become more arid, there is evidence that in a cold world, the reverse had occurred—that is, rainfall then was less strongly focused on the tropics, providing more rainfall to the extra-tropical dry lands such as the Lahontan and Bonneville basins (see figure 6.2).

So Broecker was off and running to test this hypothesis. There are many other closed basin lakes elsewhere in the world's dry lands. He asked himself what picture would emerge if evidence from all of them were to be assembled. Many questions immediately came to mind. One by one, each had to be answered.

High on the list was the need to clarify the role of North America's large glacial-age Laurentide ice sheet. Larry Benson, who has spent much of his career reconstructing the history of Lake Lahontan, had proposed that a shift in the position of the Jet Stream induced by this huge ice sheet was responsible for bringing more precipitation to Lahontan's drainage basin. If so, then one might expect that if glacial proximity rather than a colder temperature per se were the driver, the closed basin lakes at 40°S—that is, glacial-age lakes in the Southern Hemisphere where there was no large ice sheet—should have responded differently. But did they?

In the late 1980s, Broecker had visited Lago Cari Laufquen located at 40°S latitude in the dry lands of Argentina. He knew that unpublished reconnaissance radiocarbon ages obtained by Scott Stine suggested that the lake was far larger during the time of peak glaciation than it is now. So he organized a field trip to get more samples for dating. In October 2007, Scott Stine guided University of Arizona's Jay Quade, his graduate student, Alyson Cartwright, and Broecker to key locales he had previously sampled. When some months later Jay and Alyson obtained ^{14}C ages on the new samples, they confirmed Stine's preliminary results. Cari Laufquen was several times larger than now at the time of the last glacial maximum—same as the lakes at 40°N (see figure 6.3). These results reassured Broecker that global temperature played a significant role as the driver in creating wetter conditions at 40°N and 40°S during glacial times and was not simply a function of proximity to a continental ice sheet. So far, so good for Held's hypothesis.

Figure 6.2
Map showing the extent of Lake Lahontan at the time of the last glacial maximum, at the time of the Big Wet and during the late Holocene. The insert summarizes radiocarbon dates on paleoshorelines.

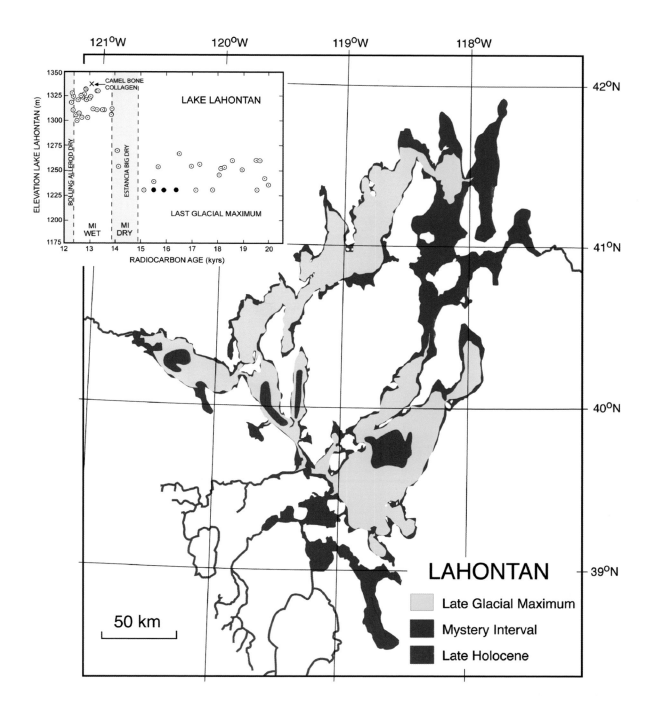

Inset figure labels:

ELEVATION LAKE LAHONTAN (m)

1350
1325
1300
1275
1250
1225
1200
1175

12 13 14 15 16 17 18 19 20

RADIOCARBON AGE (kyrs)

CAMEL BONE COLLAGEN
LAKE LAHONTAN
BØLLING ALLERØD DRY
ESTANCIA BIG DRY
MI WET
MI DRY
LAST GLACIAL MAXIMUM

Map labels:

121°W   120°W   119°W   118°W

42°N
41°N
40°N
39°N

50 km

LAHONTAN

Late Glacial Maximum
Mystery Interval
Late Holocene

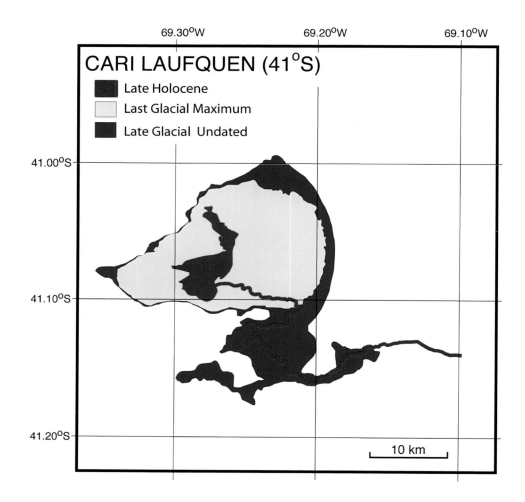

CARI LAUFQUEN (41°S)
- ■ Late Holocene
- □ Last Glacial Maximum
- ■ Late Glacial Undated

Figure 6.3
Map showing the extent of Lago Cari Laufquen at the time of the last glacial maximum, at its highest late Quaternary level (as yet undated) and at present.

Hints from the Medieval Warm

Although there was no other time during the last 100,000 years when Earth was demonstratively warmer than now, during the so-called Medieval Warm period (950 to 1300) at least the northern polar region appears to have been a bit warmer. This time period was made famous by the Norse colony in Greenland when the somewhat warmer and longer summers allowed Erik the Red and his successors to grow enough fodder to feed their livestock and hence themselves, as described in the previous chapter.

During the Medieval Warm time period, the Great Basin in the American West experienced two century-long droughts considerably more severe than any one of the five- or so-year-duration droughts of the last 150 years. Scott Stine documented these droughts based on evidence found in tree stumps now under water. Stine realized that, since these trees would have been killed by even a yearlong submergence of their roots, the water bodies now covering these underwater stumps must have been dry during the trees' lifetime. By counting growth rings, Stine could determine the duration of the droughts, which allowed the trees to grow, and by ^{14}C dating he could document when the droughts occurred.

One of Stine's research sites was the West Walker River, which feeds Walker Lake (another of Lake Lahontan's present-day remnants). During a particularly dry August, while driving along a stretch of highway 395, which parallels the river through a steep-sided valley, Stine spotted several tree stumps sticking out of the water (see figure 6.4). One turned out to have 120 rings and was radiocarbon-dated to be late Medieval Warm in age. During this entire period of the tree's growth, the river must have been dry. Starved of its river input, as a result, Walker Lake must have evaporated to a fraction of its predrought size.

Broecker spoke with Scott Stine to find out how much lakes in the Great Basin had shrunk during the time of his medieval droughts and learned that he had recently obtained an estimate for Walker Lake. He told him that starved of its river input by diversions for agricultural use, the elevation of the lake's surface had dropped 50 meters below its level in the late 1800s. He had recently walked the exposed shoreline and found stumps of small trees. Radiocarbon dates showed them to correspond in age to those for one of his century-duration droughts. At the time these trees grew, the lake was no more than half as big as it was in the late 1800s. Assuming that during this drought the reduction in size for other remnants of Lake Lahontan was also a factor of two, then 15,500 years ago, when Lake Walker reached its maximum size, it was at least ten times larger than during the time of the Medieval drought.

To achieve even an eight-fold reduction in size would require a decline by a factor of two in precipitation: a decline by a factor of two in the fraction of precipitation reaching the rivers that feed the lake and an increase by a factor of two in the rate of evaporation from the lake surface ($2 \times 2 \times 2 = 8$). Hence,

Figure 6.4
These stumps sticking up out of the West Walker River grew during a drought when river levels in this enclosed basin were much lower. (Photo courtesy of Scott Stine)

not only were these the type of changes postulated by Held, but their magnitude was extremely large.

Another observation makes things even more interesting. As evidenced by the tufa that Broecker collected in 1954 when Phil Orr took him to Pyramid Lake's Anaho Island, Lake Lahontan's largest size was not reached during the last glacial maximum, but rather during the period of deglaciation.

It turns out that this peak in the size of Lake Lahontan occurred during one of four major back-and-forth swings in the location of Earth's precipitation belts. These swings involved an *antiphasing,* that is a mirror-like pattern reversal, between the strength of rain in the Asian monsoons and precipitation in the Great Basin. These changes in rainfall appear to have been the result of shifts in the position of the thermal equator, which, as discussed in chapter 3, were driven by the extent of sea ice in the northern Atlantic. Thus, this fourfold oscillation in precipitation appears to have been driven by the appearance and disappearance of sea-ice cover around Greenland. During the first and third of the four, the Great Basin lakes were smaller and during the second and fourth, they were larger.

In order to decipher the record of these precipitation swings, it is necessary to disentangle them from the impact of changes in summer insolation related to the precession of Earth's spin axis. Because these 20,000-year Milankovitch cycles (see chapter 1) alter the summer warming of the continents, they have a strong impact on the strength of the monsoons. As our goal in these paleostudies is to understand climate change in the near future, we need not concern ourselves with the impacts of precession for there will be little change in Earth's orbit during the next century or two.

A Climate Signal from Chinese Caves

Another important climate record has recently been discovered in Chinese caves, revealing changes in the strength of the Asian monsoon rains. Scientists at the University of Minnesota, led by Larry Edwards and Hai Cheng, have demonstrated that the oxygen isotopic composition of stalagmite calcite beautifully tracks the strength of the monsoon rains. The stalagmite record the Minnesota group has painstakingly assembled reveals that the strength of China's monsoons has slavishly followed orbitally induced summer insolation cycles over the last 300,000 years (see figure 6.5). It is clear that the greater the proportion of the annual solar heating occurring during the summer months, the stronger the Asian monsoons. This finding is what we would expect since heating of air masses over land in southern Asia draws in humid air from off the ocean generating the summer rains.

But this is not our interest. Instead we focus our attention on the millennial time-scale deviations from the smooth precessional envelop and, in particular, those which occurred during the 6 thousand-year period of deglaciation—17.5 to 11.4 thousand years ago. As can be seen, there are two prominent ^{18}O maxima during this time interval. They represent times when the monsoon rain contribution to the cave water underwent major decreases.[1] Edwards and Cheng refer to them as weak monsoon intervals. The more prominent of these two weak monsoon intervals extends from 16.1 to 14.5 thousand calendar years as determined by the Minnesota group's amazingly precise ^{230}Th-^{234}U dating method. When translated to the calendar age scale, the radiocarbon ages on the tufa from Lahontan's highest shoreline Broecker originally collected with Phil Orr over 50 years ago all fall within this same range. So, at the time when Lake Lahontan Basin was at its largest size, China's monsoon rainfall was at its weakest.

The second of the pair of weak monsoon events in the Chinese cave records examined by Edwards and Cheng occurred during the Younger Dryas time interval—12.7–11.4 thousand years ago. Although this time interval was not a prominent wet interval at Lake Lahontan, it was a minor one flanked on both sides by intervals of aridity—one corresponding in age to the warm Bølling–Allerød period and the other to the warm early Holocene following deglaciation. And, of course, the very dry Bølling–Allerød time interval in the Great Basin corresponds to the time interval separating China's two weak-monsoon episodes.

The fourth of these precipitation swings occurred during the first 1,400 years of the Mystery Interval—17.5 to 16.1 thousand years ago. George Denton has suggested the term Mystery Interval for the time period between the most recent Heinrich event and the onset of the Bølling–Allerød warm period (i.e., 17.5 thousand years ago to 14.5 thousand years ago). This interval, which constitutes

1. Monsoon rains in China are observed to be more deficient in heavy oxygen (^{18}O) than precipitation occurring at other times of year. This deficiency largely is related to what is referred to as the amount effect. Rains that form in very strong updrafts greatly deplete the air mass of its moisture content and as a consequence have more negative δ^{18}O values.

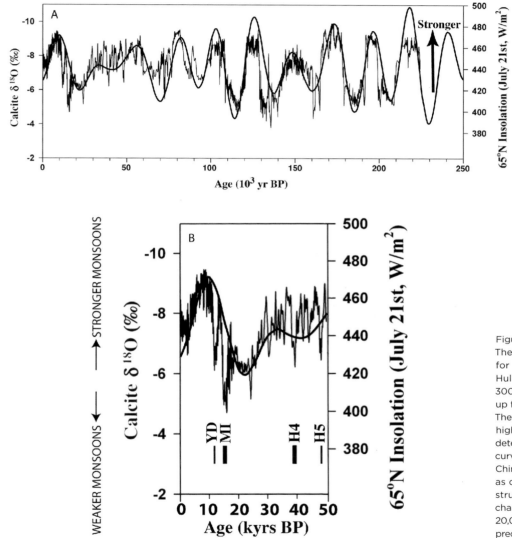

Figure 6.5
The oxygen isotope record for stalagmites from China's Hulu Cave: a) for the last 300,000 years and b) a blow up for the last 25,000 years. The age scale is based on highly precise ^{230}Th-^{234}U age determinations. The smooth curve is the solar insolation in China for the month of July as calculated from a reconstruction of Earth's orbital characteristics. The prominent 20,000-year period reflects the precession of Earth's spin axis.

the onset of the most recent glacial termination, is arguably the most fundamental climate shift of the last 100,000-year glacial cycle.

Bruce Allen and Roger Anderson as part of their study of the sediments of New Mexico's now dry Lake Estancia at 35°N discovered the existence of an episode of dessication, which they refer to as the Big Dry. They took advantage of unique exposures of the sediments deposited during peak glacial time and during the deglacial period that followed—unique, because nowhere else in the dry lands of North America does such a sequence of sediments exist. Not only is it extensive and complete, but also it contains an abundance of tiny ostracod shells, which are used to obtain a detailed radiocarbon chronology.

What Allen and Anderson found had not been seen elsewhere. The deposition of sediment in a deep and not too salty lake was interrupted by an interval of near desiccation represented by gypsum-rich sediment that contained only ostracods adapted to very salty water. During this Great Basin period of aridity, China's monsoons were "normal" in strength. In recognition of Anderson's Big Dry, it is perhaps appropriate to refer to the second half of the Mystery Interval when Lake Lahontan reached its largest extent as the Big Wet.

Sequences consistent with this Asian monsoon—Great Basin precipitation flip flop show up in records from elsewhere in the world. Most complete among these is that from the Israel-Jordan rift valley. The present-day Dead Sea and its upstream companion Lake Kinneret (the biblical Sea of Galilee) were during glacial time joined into a single large water body referred to as Lake Lisan. This lake achieved its maximum size during the last glacial maximum and again during the time of the second half of the Mystery Interval (i.e., the Big Wet). As is the case for the lakes in North America's Great Basin, Lake Lisan was intermediate in size during the Younger Dryas. It experienced major shrinkages during the times of Anderson's Big Dry and the Bølling–Allerød. In other words, its record nicely matches that for the Great Basin of western North America.

Although far less complete, we do have a snippet of record for Lake Victoria, which matches that for the Asian monsoons. Today this equatorial African lake overflows into one of the branches of the Nile River. But, based on four deep-lake piston cores, which terminate in what the University of Minnesota-Duluth's Tom Johnson has shown to be a soil horizon, this lake was once completely dry. Radiocarbon ages on limnic, or lake-deposited, organic matter from the sediment immediately overlying this soil document that the lake came back into existence early in Bølling–Allerød time. So, at least during this brief time interval, rainfall in Lake Victoria's basin followed the rhythm of the Asian monsoons.

Evidence from the Antarctic Ice Sheet

Tying all this together is an amazing record obtained by Jeff Severinghaus from Antarctica's Taylor Dome ice core. As this book centers on abrupt climate changes first discovered in Greenland, it should be stated that Severinghaus could, just as well, have carried out this study on any one of Greenland's ice cores, for it involves measurements on the ratio of ^{18}O to ^{16}O in the O_2 gas trapped in air bubbles. Winds mix the air around Earth far more rapidly than this ratio changes. Hence the results would be identical no matter which ice core was used.

The ^{18}O to ^{16}O ratio in atmospheric O_2 changes with time because its offset (averaging about 2.4 percent richer in ^{18}O) from ocean water does not remain quite constant. The origin of this offset, the so-called Dole effect, is well understood. It involves a slight preference during respiration for isotopically light O_2 ($^{16}O^{16}O$) over isotopically heavy O_2 ($^{18}O^{16}O$). Hence it is akin to the slight preference for light CO_2 ($^{12}CO_2$) over heavy CO_2 ($^{13}CO_2$) that occurs during photosynthesis. As a consequence, at steady state the ratio of $^{18}O^{16}O$ to $^{16}O^{16}O$ in atmospheric oxygen is slightly larger than the ratio in seawater—the mother lode for environmental oxygen.

What is of interest to us is that the magnitude of this offset undergoes small shifts with changing climate. While the reason for these shifts is not fully understood, they very likely involve terrestrial rather than marine plant matter. This being the case, as most of the world's land is in the Northern Hemisphere and so many of these plants are nourished by monsoonal rain, it might be expected that these small shifts in offset might follow the variations in the strength of monsoon rainfall.

Jeff Severinghaus was able to show that this is indeed the case. By making an extremely detailed and precise series of isotopic measurements on O_2 trapped in Taylor Dome ice, he was able to tease out an amazing record. Amazing, because what appear to be tiny blemishes in an otherwise smooth record beautifully match the oxygen isotope record for Chinese monsoons (see figure 6.6).

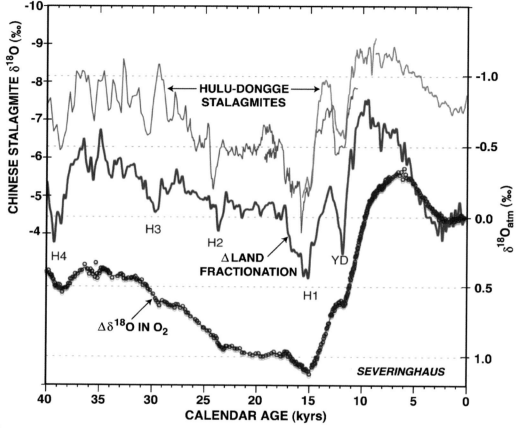

Figure 6.6
Comparison of the ^{18}O to ^{16}O record from Chinese stalagmites with that for the changing offset in ^{18}O to ^{16}O record for atmospheric O_2 (bold red curve) and the offset in the respiration fractionation factor deconvolved from the red curve (purple curve). As can be seen, the strength of the Chinese monsoons parallels the response of the entire Northern Hemisphere terrestrial biosphere to changes in Northern Hemisphere hydrologic conditions.

To demonstrate this, Severinghaus had to take into account the 1000-year turnover time of O_2 in our atmosphere. Because the time interval between hydrologic shifts averages about 1,500 years, each interval is not long enough to permit steady state to be achieved. Rather, for each shift in the hydrologic cycle, the sense of the drift in the isotopic ratio changed. On the graph in figure 6.6, at the end of the Great Basin's Big Wet, it drifted downward. Then, during the time of the Bølling–Allerød, it turned back upward and finally, during the time of the Younger Dryas, it flattened out.

In order to recover the actual changes in the respiration isotope fractionation factor necessary to create his record, Severinghaus carried out what geo-
physicists refer to as a mathematical deconvolution. As can be seen, once this
was accomplished, the results bore an uncanny similarity to the $^{18}O^{16}O$ record
in China's Hulu Cave stalagmite. Not only do the weak monsoon events associated with the Big Wet and the Younger Dryas show up, but similar ones following Heinrich events 4, 3, and 2 also occur. We can even see the smaller features
related to the Dansgaard–Oeschger events. Somehow the entire Northern Hemisphere biosphere appears to have undergone changes in precise harmony with
the strength of China's monsoons.

Evidence from South America

One more record must be mentioned before we turn our attention to the ice surrounding Greenland. There is a closed basin lake that occupies the southern part
of Bolivia's Altiplano. Everyone has heard of Lake Titicaca, which occupies the
northern part of this high Andean plateau. The southern half of the Altiplano is
largely a flat and dry plain, containing only one small lake called Poopó. Many
of us have seen the James Bond thriller *Quantum of Solace,* whose final scene
was filmed on this vast vegetation-free desert. This immense plane is the bed
of a once large lake. Age determinations on tufas from its shorelines, carried
out by Jay Quade and his University of Arizona colleagues, reveal a history
identical to that for Lake Lahontan. A far larger lake than Lake Poopó existed
during the time of the glacial maximum. And an even larger lake was present
during the time of Lahontan's Big Wet. As the Altiplano lies just beyond the
reach of the Amazonian rain belt, a southward shift in the thermal equator
would account for this extra rainfall that created this lake on the Southern
Altiplano (see figure 6.7). And, of course, the driver for these great climate
changes was the excess sea ice in the northern Atlantic following a shutdown
of the Great Ocean Conveyor Belt.

These findings raise the question of what orchestrated the back and forth
shift in Earth's precipitation belts. An appealing scenario would be to call on two
Heinrich armadas, one at 16.1 thousand years ago at the onset of the Big Wet

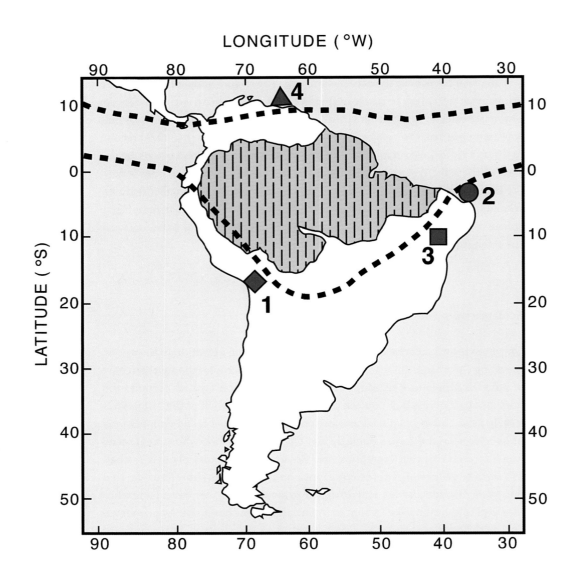

and the other at 12.7 thousand years ago at the onset of the Younger Dryas, to initiate the caps of fresh water required for the formation of extensive northern Atlantic sea ice. Perhaps periods of warmth during the Big Dry and the Bølling–Allerød were responsible for creating instabilities, which caused these massive collapses of the ice lobe centered over Hudson Bay. These intervals of sea ice cover would have been terminated by spontaneous rejuvenation of conveyor circulation. It sounds sensible, but as with many ideas concerning the conveyor's impacts, it may well bite the dust.

One other puzzle has yet to be unraveled. It has to do with an anomaly in the timing of the retreat of the western portion of the Laurentide ice sheet. To the best of our knowledge all the other ice masses of glacial time began their major retreat during the time of the initial CO_2 rise (17.5 to 14.5 thousand years ago). Included are that portion of the Laurentide ice sheet east of the Mississippi River, the Scandinavian ice sheet, the mountain glaciers of the European Alps, the Southern Andes, and the New Zealand Alps. Yet that portion of the Laurentide ice sheet to the west of the Mississippi River reached its maximum during the Big Wet period (i.e., two thousand or so years after all the other ice masses had begun a full-scale retreat). This has something important to tell us, but so far that "something" eludes us.

So, it appears that the record of paleoprecipitation does lend support to Held's prediction. This record also tells us that shifts in the thermal equator redistribute rainfall. This latter observation has an important bearing on what will happen as a result of global warming, for models make a firm prediction that because of its greater land area the Northern Hemisphere will warm up faster than the Southern Hemisphere. Hence, perhaps the thermal equator and its associated rain belts are due for a northward shift.

Figure 6.7
Map of today's Amazonia and the location of sites where evidence has been obtained suggesting a southward shift during the Mystery Interval and Younger Dryas: 1) enlargement of the Altiplano lakes, 2) enhanced river runoff, 3) stalagmite growth in otherwise dry caves and 4) reduction of river runoff into the Caribbean Sea.

CARBON DIOXIDE AND THE FATE OF THE GREENLAND ICE SHEET

The Carbon Dioxide Question

In the previous chapters, we described some of what we know about the paleoclimatic history of Greenland and its relation to the history of the planet. But you might wonder, what does this have to tell us about the future? A lot, it turns out. To see this, though, we need just a little additional background.

We humans are rapidly converting fossil fuels into carbon dioxide, releasing the solar energy stored in the coal, oil, and natural gas in the rocks beneath us about a million times faster than nature saved it. CO_2 is rising, contributing to observed warming. With a lot of fossil fuel left to burn, there is high scientific confidence that the world will warm a lot in the future unless we change our behavior.

Now, some readers are nodding their heads in agreement with the previous paragraph, while others are shaking their heads in disagreement. "You don't KNOW that," we can almost hear some of you thinking, "Because there are experts on both sides of the issue."

Actually, no, there are not experts on both sides of the issue. Instead there are many issues; and although there is essentially no disagreement about some of these issues, there is plenty of room for disagreement about some of the most important ones. Let's drill down through this a bit.

First of all, are we indeed raising CO_2 by burning fossil fuels? Yes. Although you can find arguments in the blogosphere about almost everything including

Figure 7.1
A view looking down from the Julianehab Ice Cap over a nunatak into Tasermuit Fjord in South Greenland. Note to the upper right a flat surface above the fjord wall with a sea of frost-shattered blocks of rock, known as a *felsenmeer*. (Photo by Gary Comer)

whether the Earth is flat, the Earth really is roughly spherical, and we are raising CO_2 levels in the atmosphere. To convince yourself, all you really need to do is a little bookkeeping. We know about how many oil tankers and coal trains there are; we know what is in oil and coal chemically; and we can see the CO_2 from the burning oil and coal building up in the air and the surface ocean. Volcanoes are pathetically small sources in comparison—roughly 1–2 percent of the human source—and have not done anything of importance to global climate recently. The recent Icelandic volcano, for example, had no global effect. And, geochemical tracers exclude the volcanoes as a possible source. A volcano makes most of its CO_2 by melting things that already contain CO_2, and so a volcano that raises CO_2 doesn't lower oxygen, whereas our burning of fossil fuels to make CO_2 uses oxygen. The measured drop in the air's oxygen content is just what you'd expect to explain the rise in CO_2 if it comes from burning fossil fuels. You can dredge to the far-fringing edges of discussion to find someone who will argue about this, but there is really nothing to argue about—we are raising CO_2.

Next, does more CO_2 contribute to warming? Yes. The basic physics of CO_2 as a greenhouse gas have been known for more than a century. During and after World War II, the military became interested in the role of CO_2 and other gases in many things, such as blocking the view of heat-seeking missiles, and the resulting careful research on electromagnetic radiation in the air removed any doubt about the role of CO_2. It may be useful to think of the warming effect of CO_2 in the same way as the effect of hiring additional help in a company— each new additional employee partially duplicates the role of those already there, but not entirely, so new increases in the payroll or in CO_2 have an effect, but less effect than the earlier increases. This diminishing-returns aspect of CO_2 is fully included in projections of the future, just as businesses recognize the diminishing returns of new hires. Again, a few people with scientific initials after their names will gladly argue about this, but there is no sensible argument to be had—more CO_2 tends to warm the planet.

So, we are raising CO_2 and having a warming influence. Will this warming influence overcome natural variability and lead to real warming, and, how much real warming will we see? Here, the discussion begins to get more complicated. The best estimate is that, if you start from a stable climate, then double the CO_2 concentration in the atmosphere and let the climate adjust, you will have

Figure 7.2
Currents in Scoresby Sound carry large floating icebergs that have calved off of rapidly flowing tidewater glaciers. These large, flat ones are called tabular bergs. They floated in the fjord while still attached to Daugaard-Jensen Glacier, before breaking free. Photo by Philip Conkling)

caused a warming of between about 2 and 4.5°C (just under 4° to just over 8°F), with a most-likely value of about 3°C (just over 5°F).

The direct effect of the doubled CO_2 is a bit over 1°C, and the rest is what we call feedbacks. For example, snow and ice reflect the sun's energy efficiently, bouncing the light back to space without allowing the light to warm Earth. Warm the planet through CO_2 or any other cause, the warmer conditions melt some of the snow and ice, more sunlight is absorbed, and the planet warms a little more. Warming causes some changes that in turn make the warming bigger—we call these positive feedbacks—and warming causes other changes that in turn make the warming smaller—we call these negative feedbacks.

Our understanding of the climate system is that the positive feedbacks win over the time intervals of most interest to humans, so that temperature changes over years to centuries are larger than you would expect from the size of the initial cause. This is true whether the initial cause is more CO_2, or a brighter sun, or a giant space-alien heat-ray, or anything else you might think of to cause warming. But, because the feedbacks involve evaporation of water, formation of clouds, changes in snowfall and melting, changes in vegetation, and more, there is still plenty of scientific research to be done on these issues.

Notice that the uncertainties associated with this research are already included in our assessment of likely futures—when we write that doubling of CO_2 is expected to cause warming of 2 and 4.5°C, some might immediately say "Ah, but you are very uncertain, so maybe that could be zero..." Climate scientists use a range because the uncertainty described above is already included. Many more-detailed discussions of this range are possible, and in general they point to a high likelihood of warming from doubled CO_2 falling in the range of 2 and 4.5°C, with a small chance of a slightly smaller value, and a small chance of a slightly larger or even much larger value. But the important point is that the feedbacks are very unlikely to be negative overall, and because the different positive feedbacks amplify each other, the slight chance of a really big change cannot be excluded entirely.

You can find scientists, good scientists, who argue for warming near the lower end of this range, and others who argue for warming near the higher end. There is room for disagreement, but that disagreement is already included here. There is enough fossil fuel in the ground that, if we burn it rather rapidly, we can easily double the level of CO_2 that was in the air when the industrial revolution

Figure 7.3
Sea ice off Illulissat (Jakob-shavn) off the west coast of Greenland creates hazardous conditions for small boats. (Photo by Gary Comer)

began. We can probably quadruple it, and we might even come close to octupling it. Some experts even think that we can be sufficiently clever in finding fossil fuels, including sea-floor methane, that we can go far past octupling.

Each doubling of CO_2 has roughly the same effect on temperature, so we can think of temperature increases of three, six, or nine degrees centigrade depending on how much fossil fuel there is and how much of it we burn, with the range of uncertainty perhaps between 2°C for one doubling with low sensitivity, or a whopping 13.5°C (or more) for an octupling with high sensitivity. The extremes are rather unlikely, though. You can find plenty of scientists and economists and others who will argue about the reserves of fossil fuels, and whether we will really burn them all, and the best estimates are a little more nuanced than we gave here. For example, we won't get quite as much warming as presented here if the CO_2 concentration in the air starts down before the ocean has time to finish warming, but we might get more warming than described here if a lot of carbon stored in the permafrost and sea floor is released by the warming.

Is this warming bad? Is it a crisis? What should we do about it? At this point, there is lots of room for debate, and you will find experts on many sides of the multiple issues. We as authors have no intention of answering these questions for you, and we probably would come up with a range of responses just among ourselves. But, a few things come out pretty clearly across a very broad range of experts.

We have adapted our cultures and economies for the climate we have, so in the short term, pretty much any change is costly. For global changes, if we burn most or all of the fossil fuels, the combined size and speed of the coming changes will be large compared to natural shifts. The world has changed hugely in the past, but getting from the ice-free world of the dinosaurs to our world took something like 100 million years, not 100 years. The meteorite that killed the dinosaurs 65 million years ago had a huge global impact rapidly, but the only events with global reach that caused persistent changes at a rate similar to or faster than what we may do are the abrupt changes recorded in the Greenland ice cores. Significantly, those abrupt changes had very large effects on temperature, rainfall, winds, and plant and animal growth, all around the globe—big climate changes really do have big impacts on the living things of the planet.

Figure 7.4
Turbid meltwater carries silt eroded by the glacier into the edge of Melville Bay, on the northwest coast of Greenland. The ice flows more slowly, spreading down from the two-mile-thick center of the ice sheet. (Photo by Gary Comer)

The saga of the Norse is important in thinking about the future. Although much of the world—the big belt around the equator—seems a bit warm for human comfort, a bit of the world—the small caps at the poles—is on the cold side. Cooling really stressed the Norse, and when combined with other economic issues, it forced them out. Similar stories can be told for many other parts of the world, for different people and different climate changes; when the prevailing conditions shifted, people were pushed, with droughts in particular causing civilization-scale disasters. Notably, however, people with enough knowledge and enough energy sources and enough bulldozers and tractors and pumps and such are less influenced than were our ancestors. A cooling comparable to the Little Ice Age that forced the Norse away would not similarly defeat their modern relations living in Greenland.

In general, when we are faced with a looming problem, we can learn to live with it, or head it off. The best science indicates that global warming will stress our society, especially poor people living in hot places. Based on a preponderance of good science, including economic analyses, the wisest response appears to include both reduction of CO_2 emissions—through some combination of new energy sources including conservation, and capturing the CO_2 and putting it back in the ground—together with helping people handle the changes, in part by raising the standard of living of poor people. Screaming panic is not recommended (if you are reading this on a plane, you do not need to walk home), but neither is business-as-usual, which is highly likely to bring very large costs for this generation and especially for future generations.

Figure 7.5
The general warming trend in Greenland is revealed in this view of the receding glaciers in the Stauning Alps in East Greenland. The smooth regions around the remaining glaciers were ice-covered a century or so ago near the end of the Little Ice Age. These glaciers have almost no accumulation area this year, and are likely to melt away completely in the not-too-distant future. (Photo by Philip Conkling)

Sea-Level Rise

The fate of Greenland is deeply entwined with the wise path, in two additional ways: sea-level rise, and abrupt climate change. We'll say a bit about sea-level rise here, and come back to abrupt climate change in the next chapter.

We saw in chapter 2 that the ice sheet on Greenland is a two-mile thick, very wide pile of old snow, which spreads under its own weight. During the last ice age, cooling had allowed the ice sheet to grow, and the warming from the end of the ice age forced retreat. For the last 10,000 years or so, changes

Figure 7.6
Melting of icebergs does not cause sea-level rise, but warming increases the flow of ice from non-floating glaciers to make icebergs, which does cause sea level to rise. Here, some of those bergs collapse into brash ice and melt in Northwest Fjord in Scoresby Sound. (Photo by Philip Conkling)

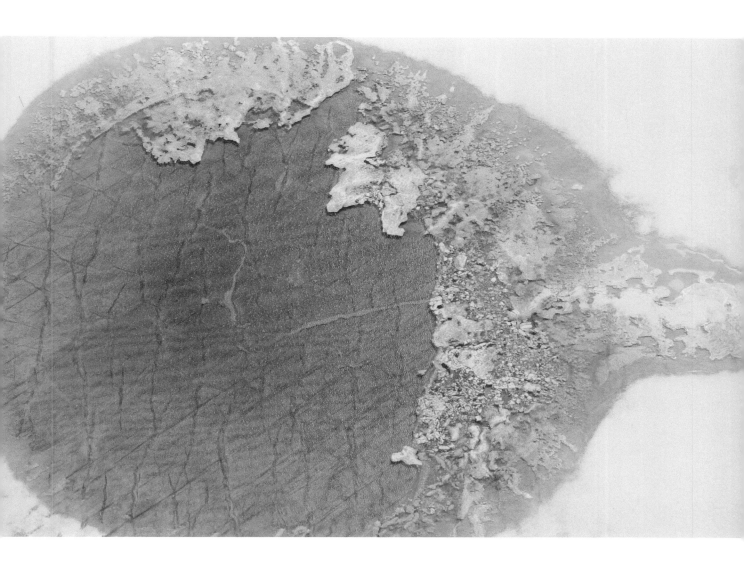

Figure 7.7
Melt water on the surface of an ice cap can accumulate in low places, creating gloriously beautiful blue lakes. Here, the surface of the lake refroze, ringing the right-hand side and upper part of the lake. Waves are visible on the rest of the lake, above the traces of once-open crevasses that have since been squeezed closed again, and that run in different directions. (Photo by Gary Comer)

in the ice sheet have not been nearly so large, with small warmings causing slight shrinkage and small coolings causing slight growth of the ice. Looking further back, warmer times caused notable shrinkage of the ice sheet, with a few degrees of warming (and some uncertainty about just how many degrees) sufficient to greatly reduce or eliminate the ice.

Over the last few decades, there has been a general warming trend in Greenland, but with plenty of bumps on the way. The ice sheet has tracked the climate closely—warming brings more snowfall to the cold central regions of Greenland, thickening the ice there, but it brings more melting to the warmer coastal regions. The increase in melting with warming beats the increase in snowfall, so the ice sheet shrinks with warming, raising sea level. Furthermore, warming increases the flow of ice from the land to make icebergs, again raising sea level. Finally, warming of the ocean causes thermal expansion as the water of the oceans expands, also raising sea level.

Two things are especially important in the acceleration of ice flow with warming. First, the bedrock under Greenland is bumpy. When the flowing ice encounters a big bump in the bed, some ice flows around it but some flows over, pushed by the ice behind, and the general downhill trend from the center of the ice sheet to the coast may be locally reversed. (Somewhat similarly, a kayaker in a wild river may surf upward on a wave over a big boulder.) Melt water on the surface of the glacier can accumulate in the low places, creating gloriously beautiful, deep-blue lakes. When ice flow opens a crevasse under a lake, the weight of the water wedges the ice open more, and the water cascades down to the bed. The flow can be larger than Niagara Falls for an hour or so, and the water tends to lubricate the bed and make the ice go faster.

More important is that in many places, the ice that reaches the sea doesn't immediately break off to form icebergs, but instead remains attached to the ice sheet even though it is floating in a fjord. Friction between these floating extensions—called *ice shelves*—and the adjacent rock helps hold back the ice sheet behind. When warming causes melting of the ice shelf, from above or below, the friction is reduced, the flow of the ice behind accelerates, and this accelerated flow of the nonfloating ice behind contributes to sea-level rise. In the 1980s, Jakobshavn Glacier on the west coast of Greenland still had an ice shelf, and it just may have been the fastest glacier on the planet. Then, as warming attacked and removed its ice shelf, the flow velocity approximately doubled.

Figure 7.8
In some places, ice floats in the sea while still attached to the glaciers behind, forming ice shelves. Ice shelves generally are restrained by friction with adjacent rocks, as shown here at the north side of the Stauning Alps, and this restraint slows the flow of non-floating ice into the floating ice shelves. When warming of air or ocean weakens or removes ice shelves, the loss of friction allows the non-floating ice to flow faster into the sea, raising sea level. (Photo by Philip Conkling)

Figure 7.9
Massive icebergs from the Sermeq Kujalleq Glacier near Ililussat on Greenland's west coast are carried by ocean currents around Baffin Bay and into the North Atlantic before they melt. Note the gull atop the right-hand berg, for scale. (Photo by Gary Comer)

Figure 7.10
A Greenlandic fishing vessel passes by a massive iceberg north of Ilulissat. (Photo by
Gary Comer)

There are still places in Greenland where warming can attack ice shelves. Sea ice in fjords can act like a thin ice shelf in winter, holding back the ice, and warming can also attack this sea ice, in Jakobshavn and elsewhere. So, more warming can cause more ice-flow acceleration contributing to sea-level rise.

The Greenland ice sheet now is high and cold in the middle, lower and warmer on the edges. If too much ice leaves, the center will be lowered, getting warmer. At some point, thinning from warming will cause enough more warming that the ice sheet won't survive. The competition between thinning from warming and thickening from more snowfall makes this a difficult tipping point to predict, but the history of the ice sheet agrees with our understanding of the physics that there is such a tipping point. We don't believe that the ice sheet could fully disintegrate faster than many centuries, but we might cause enough warming within a few decades to cross the threshold leading to ice-sheet loss. Then, unless we find a way to pull the temperature down rapidly, the ice sheet would be on the way to disappearing.

The ice sheet is big—7.3 meters or 23 feet of sea level is tied up in Greenland ice. Melt all of Greenland's ice, and the world's oceans will rise a lot. In comparison, the deepest water in New Orleans after hurricane Katrina breached the levees was about 20 feet; the water was pumped out after a couple of weeks, but the damage was enormous. The fate of Greenland has the possibility of putting all of the low-lying parts of the big coastal cities into such a predicament. Because Antarctica's ice is much bigger than Greenland's, and Antarctica now also seems to be losing mass from warming, even bigger problems are possible from melting ice.

OUT OF THE ICE

The Lessons from Greenland

The public debate about climate change has been treated in the press as having two sides—the science showing that our business-as-usual actions will change the climate in ways that on average hurt us, and the opposition arguing that things might end up better than that projection. As noted in the previous chapter, such a two-sided view has always been overly simplistic—there are many more sides; in particular, the possibility that things might end up a lot worse. Our research in Greenland is part of a larger body of emerging science on abrupt climate change, emphasizing the lop-sidedness of the uncertainties. The future could be better than the mainstream scientific projections, but "a little better" is balanced against "a little worse" or "a lot worse," as reviewed below.

Climatic Confirmation

If the sun brightens, or Earth reflects less sunlight, or Earth's greenhouse gases rise, the planet warms. Such changes have come and gone naturally, and they will again.

If we humans continue with business as usual, however, we are likely to cause changes for the whole Earth of a combined size and speed not seen in a very long time, if ever. The CO_2 production in the United States is equal to approximately 20 tons per person per year—each of us responsible for roughly 250 times our own weight in CO_2 each year, or roughly 20,000 times our own weight over the course of our lifetime at current rates. And, with much of the rest of the world either fairly close to this level of carbon-dioxide production or else trying to get there, that's a lot of CO_2. Much of that CO_2 goes into the ocean, and some goes into trees that grow a bit better with the extra CO_2, but we are still raising the CO_2 concentration in the atmosphere, and the atmospheric level will remain elevated centuries and even millennia from now.

The mere volume of CO_2 doesn't tell us how important it is—if we released that much water vapor, it would rain out in a week or two, with no one seriously noticing the difference. But if we were to vent even a fraction of that amount of many poisons into our cities, such a release would cause very different results. Burning most of the fossil fuels and leaving the CO_2 in Earth's natural systems will warm the world a lot, with large impacts on ecosystems and economies.

Our tour of Greenland has helped provide perspective on this. The ice core records give us many insights into the causes of climate change, and together with moraines and other indicators, the ice in Greenland has revealed how the climate responded to those causes. Understanding these causes and effects allows us to say a lot of common-sense things about the climate, and it allows quantitative testing of the climate models that are used to project future changes.

Our understanding of greenhouse warming rests primarily on straightforward physics, not on correlations. And the physical understanding of energy transfers and greenhouse gases makes sense of a climate history that is otherwise inexplicable.

For example, the globally large and near-synchronous climate changes of the ice-age cycles were caused by features of Earth's orbit that had almost no effect on the total sunshine reaching the planet. Some places cooled while receiving more sunshine, and ice grew in places where extra summer sunshine favored melting. But as described in chapter 1, CO_2 dropped with the ice age and the effects of that CO_2 decline successfully explain the size and global reach of the ice age cooling. No other successful explanation has ever been advanced.

Figure 8.1
Field team at entrance to Northwest Fjord, Scoresby Sound; from left, Telford Allen, pilot; George Denton, geologist; Richard Alley, glaciologist; Philip Conkling, writer. (Photo by Gary Comer)

Figure 8.2
View northwest toward the late-glacial moraine loop in the Itivdlerssuaq valley. The glacier ended where the bare till meets the tundra at the historical maximum around 1850, but it had turned the corner and extended far out onto the main valley floor during late glacial (probably Younger Dryas) time. This is perhaps the most spectacular late-glacial moraine in southern Greenland, which can help reveal the causes of climate change there. (Figure 3.3 also shows this moraine, but looking in from the left side of this picture.) (Photo by Gary Comer)

We can look at the ice age to estimate the climate-sensitivity—the warming from raising CO_2 by a given amount, or the cooling from lowering CO_2 by that same amount. Some of the ice-age cooling was provided by the reflectivity of the large ice sheets, some from the sun-blocking ability of the extra ice-age dust, and some from the effects of the changed CO_2 and other greenhouse gases. If you account for the cooling this way—some from reflectivity of the ice, some from dust, some from the other things we know about, and the rest from CO_2—then the CO_2 effect agrees closely with the estimates from our modern models and from other comparisons to paleoclimatic data as given in the previous chapter, and is usually expressed as ~3°C (5°F) warming for doubled CO_2 (although CO_2 didn't quite double in going from the ice-age to the natural levels before the industrial revolution).

However, such an accounting is in some ways cheating, and it underestimates the effect of the CO_2. The ice sheets would not have grown as big without the cooling from the drop in CO_2. In turn, the ice sheets would not have been as reflective if the CO_2 had not dropped. Thus, the CO_2 really caused some of the cooling that we credited to ice-sheet and dust reflectivity in the previous paragraph, and so the ice-age changes indicate that CO_2 is even more influential than indicated by our models. However, our models may not be too bad if you are only worried about the next decades, because really big changes in ice sheets take centuries or longer to occur.

The ice-core and moraine records provide many other insights, too. Big volcanic eruptions tend to cool the climate a little for a year or two, and a cluster of eruptions can result in cooling that persists over many years or even a few decades, but volcanoes have trouble getting organized. The sun does vary, and the climate responds, but we are fortunate that there is no sign of big, fast changes in the sun. Space does not send enough dust our way to matter much. The climate has not tracked the changes in cosmic rays, so they cannot matter much either.

Thus, we have a pretty good understanding of the things that control Earth's average temperature. We can estimate the size and rate of the natural changes including the slow but predictable orbital cycles and the small but less-predictable volcanic eruptions and changes in the sun.

Figure 8.3
Other indicators of climate change include the maximum extent of ice coverage over mountain peaks, such as these over Milne Land. As shown in Figure 3-9, the jagged peaks were not overrun by the ice sheet, but the rounded bedrock near the camera and down the flanks of the peaks was once beneath glaciers. The peaks shown here rise almost a mile and a half above the water in Northwest Fjord, which is the deepest fjord in the northern hemisphere, plunging almost a mile below the surface, making this area one of the steepest, highest-relief places on the planet. (Photo by Gary Comer)

Living Locally

No one really lives in Earth's average climate, so no one really cares about it. And here things get more interesting for most of us.

The abrupt climate changes of the past, such as the jumps into and out of the Younger Dryas event and the other Dansgaard–Oeschger oscillations we have discussed, did not do much to the global average temperature, with cooling at one end and warming at the other end of the oceanic conveyor-belt seesaw. But almost everyone would have noticed that their climate changed during the jumps, and some people would have noticed big changes.

When too much fresh water puddled in the North Atlantic—perhaps from the sudden draining of glacial Lake Agassiz, or for other reasons—the wintertime freezing was accompanied by a southward shift of the tropical rain belts across Africa and the Americas with reduced rain in the monsoonal regions of Asia, among other changes. Were something similar to happen in our future, it would have large implications; a whole lot of people live in regions that in the past were made drier when the North Atlantic was freshened. We have more land in the north than in the south, so moving the rain belts south tends to reduce the rain for people, who live on land.

When we started working on these abrupt climate changes, we were very concerned that another such event could happen soon, as was described in chapter 4. Melting of Greenland and of additional Arctic ice, along with other changes in the far north, are contributing extra fresh water to the ocean. Simple ocean models suggest that this much water might overwhelm the conveyor circulation. Even though the warmer climate state of today in comparison with Younger Dryas time would be likely to mute the sea-ice response, very large impacts on humanity could still result.

Further research has notably eased our fears. As ocean models have become better, they have indicated less and less possibility that Greenland's melt water will arrive fast enough to truly disrupt the conveyor. The most recent assessment by the Intergovernmental Panel on Climate Change gave a greater than 90 percent chance that nothing drastic would happen to the conveyor over at least the next century. This is good news, indeed.

But, 90 percent is not 100 percent, and a small chance is not necessarily a zero chance. Furthermore, the North Atlantic is not the only place where abrupt

Figure 8.4
The melting of Greenland ice and other changes in the Arctic region around Greenland are contributing extra fresh water to the North Atlantic, including these icebergs in Scoresby Sound. (Photo by Gary Comer)

changes might be possible. We are confident that ice sheets don't just "fall into the ocean" and raise sea level, but our lack of fundamental understanding of ice-sheet behavior leaves open the possibility that we could be greatly underestimating the rate of response to warming, with potentially major implications. We are hopeful that ecosystems such as the Amazonian rain forest will prove resilient against climate changes, and will change slowly if at all, but the possibility of a sudden switch to a different ecosystem cannot be ruled out. Methane locked in seabed sediments will probably behave itself, leaking out slowly if at all, but we cannot absolutely guarantee that there will not be large and rapid releases.

Significantly, these and other possible tipping points all slant in directions that would cause harm to economies and existing ecosystems, with losers substantially outnumbering winners. And we do not see any evidence for tipping points to the other side that would suddenly jump us into a wonderful climate with winners greatly outnumbering losers. We have built our cities, societies, and economies for the climate we have, and our ecosystems are adapted to our present-day climate. Thus, big jumps tend to go in bad directions, a little like trying to adjust a watch with a hammer.

Greenland brings us face-to-face with more issues than this. The conveyor circulation has been chugging along without major interruption for more than 8,000 years, but it probably does wobble a little on the way, speeding and slowing in response to its own dynamics and in response to pushes from the sun and volcanoes. Scientific research is ongoing, but it seems likely that such changes shifted the North Atlantic's currents and climate into a somewhat different pattern contributing to the slide on adjacent lands from warmer medieval times to the Little Ice Age. Furthermore, similar events probably occurred several times before, with a roughly millennial spacing.

For the Norse people, living near the economic and climatic edge in Greenland, the climate changes loomed large. Although vigorous debates still exist over the reasons that the Norse settlements failed in Greenland, climatic changes almost certainly mattered. Furthermore, climatic impacts on people were evident across Europe and beyond.

Vigorous research continues on climate patterns during even older times and for other areas, but similar patterns are emerging in many places. Droughts recorded in cave formations in China correlate with the ends of dynasties. In much the same way, droughts marked the times when Ancestral Puebloan, Mayan, and other civilizations were reorganized or fell, with evidence that the people were stressed by those droughts. It remains possible, if unproven, that some of these paleoevents were linked to the conveyor changes, and other events seem to have been related to the solar and volcanic forcings of climate and to resulting changes in the behavior of the ocean-atmosphere system. Simply turning the sun up a little or down a little does not by itself cause such big changes, but the patterns of the climate response to the changing sun can amplify those effects, often causing some places to become warmer while others grow colder; or some get drier while others become wetter.

The most consequential of the pattern changes is potentially the biggest. The powerful tropical circulation in the atmosphere, where hot air rises and drops rain in a narrow near-equatorial belt before sinking and drying in a broad subtropical belt, moves an immense amount of heat and moisture around the globe, affecting huge numbers of people. The history of climate, from Greenland as well as from the lake basins and cave formations of the world, shows that the size and position of atmospheric circulation can change rapidly. In particular, the dry zone of subtropical sinking air seems to contract with cooling and expand with warming and to shift toward the warmer side if temperature changes more in one hemisphere than in the other. The tropical circulation interacts with the tropical to subtropical monsoonal circulations, which have also responded to these large-scale climatic shifts.

Models of the future indicate that with CO_2-induced warming, the zone of rising air and heavy rains will become more vigorous but remain narrow, while the dry sinking zones become wider and drier. This dynamic will further desiccate already-dry regions while shifting some regions from wet to dry. Whether the rising zone will move a little as the planet warms up remains unclear but is an important topic for additional research, and the expected faster warming of the northern lands leads to the expectation of a shift. Such a pattern is already emerging in recent observations. The paleoclimatic record points to very large past changes in the tropical circulation, and the models on which we rely for future projections do a pretty good job of reproducing the past trends, increasing our confidence that the models are working correctly.

Unfortunately, this is not good news. Drying and expanding those already-dry zones is not helpful in a world with a lot of hungry people, especially if we

remember the immense size of the dry zones spread around Earth's midriff. The emerging picture is that climate changes often trouble humans, and especially humans who cannot zip down to the grocery and buy food from somewhere else in the world if the local crops fail. Cooling already-cold places or heating already-hot places, drying the deserts or flooding the wet places all can cause trouble for us. A whole lot of the world is already hot and dry, and we're turning the thermostat up and expanding the biggest of the dry zones.

Ongoing Options

Global warming is physics, not a hoax. And, physics tells you what is and what may be, not what ought to be done about it.

As climate scientists, we face great challenges to improve our understanding. The future of global average surface temperature is a starting point, not a destination. Understanding the future patterns of rainfall, and summer and winter temperature regimes for individual regions are all critical parts of our research agenda. So is the future of the conveyor, along with El Niño, the linkages with tropical circulation and understanding the North Atlantic Oscillation, the Atlantic Multidecadal Oscillation, and more. All of these natural phenomena are likely to respond to changes in the global average temperature, as well as to other forcing features. And, we cannot yet predict the future of these effects with high confidence. Warming melts ice, but how much ice will melt and how rapidly in Greenland and Antarctica are very poorly known, yet they are vital to coastal people all around the world.

Even without this refined scientific understanding, though, we have high scientific confidence that the world will warm, the dry zones will become drier and expand, the ice will melt, and millions and millions of people will be affected. Although warming some very cold regions may create new winners, so much of the world is warm and dry that warming and drying appear poised to bring many more losers, with some of the problems already being present.

The best estimates are that the wealthier "first world" will be able to handle what is coming without huge problems, although with notably more costs

Figure 8.5
Although vigorous debates still exist over the reasons that the Norse abandoned their Greenland settlements that they had occupied for almost 500 years, including the medieval church and farm at Hvalsey pictured here, increasingly colder climates almost certainly were a factor. (Photo by Gary Comer)

than benefits, but that the third world faces much graver issues. Back in Greenland, climate change helped drive the Norse out as the Little Ice Age arrived. But if the Norse could have met the change by flying in a bishop and a bunch of wealthy tourists on planes bringing food and warm clothing, it is hard to imagine them abandoning settlements that must have had truly spectacular scenery as the great glaciers advanced down the fjords.

Studies that couple climate and economy and ask what is the best path forward almost always choose some combination of limiting the warming and helping people deal with the changes, but the balance of these two challenges varies notably from study to study. Because unfavorable changes already have occurred in some places, there is not enough time to stop the warming or to give everyone resources before the harm arrives.

The choice of how to balance climate change and economic dislocations will become much easier if new technologies are developed to supply energy while holding down CO_2 emissions. This may involve CO_2-neutral energy sources (wind, sun, nuclear, etc.), or capturing the CO_2 from fossil fuels so it doesn't go into the air. Where science meets engineering and invention, there are great opportunities for bright people to make a big difference—and a lot of money—on future energy.

Greenland also reminds us that the science is not certain, and that this uncertainty is not good either. The models may overestimate or underestimate the coming warming a little, or underestimate it a lot, but the physics argue against a big error on the other side. With the possibility of abrupt climate changes, which have occurred repeatedly in the past, lurking in the future even if unlikely, we must recognize the possibility that we have greatly underestimated the coming damages of climate change. However, we do not find evidence that we have greatly overestimated the damages.

When the Norse occupied Greenland, they did not have modern insurance companies. Nevertheless, successful merchant traders did invest in shares in several different voyages to spread the risks against a catastrophic loss of a single ship and its cargo. Thus even the premodern Norse had a rudimentary concept of the wisdom of insurance. Tell a modern economic model attached to a climate model about our uncertainties, the existence of abrupt climate changes, and the chance of high climate sensitivity, and the model may buy insurance by

Figure 8.6
How much and how rapidly ice melts in Greenland is poorly understood, yet is of vital concern to coastal people all around the world. (Photo by Gary Comer)

Figure 8.7
The effects of a warming climate are expected to give economic "winners" and "losers," but with losers growing to greatly outnumber winners. Tourists may flock to the endangered but spectacular scenery of calving icebergs in places like Ililussat (Jakobshavn), where approximately 10 percent of the icebergs in the North Atlantic are spawned. However, much larger regions that are already warm, and the many more people who live there, are expected to be disadvantaged. (Photo by Gary Comer)

slowing down the warming more. Slower emissions probably reduce the chance of crossing a tipping point and give the scientists and engineers more time to figure out solutions to the problems.

Back from the Ice

We have had the good fortune to study the beauty of the planet in ice and mountains, lakes and seas, deserts and rainforests. We have seen the resiliency of the people in these places. And at the end of a field research season we return to the bright people of our universities and communities. In short, we get to see clearly the causes for optimism about our future.

Yet, this optimism must be tempered with concern, because success is far from guaranteed. The ice-core records confirm our understanding of the climate, and they give us great confidence in our projection that business-as-usual human fossil-fuel burning will change the climate in ways that affect us and other living things, with the effects becoming more negative as the climate changes become bigger. The great mass of ice responds to the climate, and warming is highly likely to bring melting and sea-level rise, affecting the coasts of the world. The climate's response to the abrupt changes of the past, and to the ice-age cycles, points to the likelihood that the warming of the future will cause shifts in the rain belts with more drying than wetting, affecting millions or even billions of people.

Recent assessments of abrupt climate changes have concluded that no single type of abrupt change is considered likely. We don't expect a North Atlantic conveyor shutdown, or a belching of methane, or a sudden collapse of a big ecosystem, or a sudden ice-sheet collapse, but all appear possible. Even though we know that big, fast, and bad climate effects could occur, as scientists we cannot in good conscience actually predict with current information that Earth will soon jump far to the bad side. Nevertheless, it is also true that the uncertainties are dominantly on the bad side.

Far from being an excuse for inaction, good scientific and economic analyses show that uncertainty favors action to reduce greenhouse-gas emissions,

Figure 8.8
A view of the reconstructed Medieval church built by Erik the Red's wife at Brattahlid, a farm occupied for half a millennium before changing climate contributed to the abandonment of the settlement. (Photo by Gary Comer)

Figure 8.9
The great mass of ice in Greenland and Antarctica responds to climate, and Earth's warming
is highly likely to bring melting and sea level rise, affecting the coastlines of the world.
(Photo by Gary Comer)

Figure 8.10
Gary Comer, who sponsored a decade of field research in abrupt climate change in Greenland and around the world, left a legacy of addressing the question of our place on the planet with intelligence and vigor that greatly improves our prospects for the future. (Photo by Gary Comer)

because the most-likely future under business as usual is fairly close to the good end of the range of possibilities, with most of the uncertain-but-possible outcomes on the bad end.

The lessons of abrupt climate change in Greenland and elsewhere are clear. Ignoring our environmental impact will not make it go away, but considering our place on the planet with intelligence and vigor greatly improves our prospects for the future.

BIBLIOGRAPHY

Agassiz, L. 1886. *Geological Sketches, Second Series, Ice-Period in America.* Boston: Houghton, Mifflin and Company. *Geophysical Research Letters* 28:2077–2080.

Alley, R. B., J. T. Andrews, J. Brigham-Grette, G. K. C. Clarke, K. M. Cuffey, J. J. Fitzpatrick, S. Funder, et al. 2010. History of the Greenland Ice Sheet: paleoclimatic insights. *Quaternary Science Reviews* 29:1728–1756.

Alley, R. B. 2007. Wally was right: Predictive ability of the North Atlantic "conveyor belt" hypothesis for abrupt climate change. *Annual Review of Earth and Planetary Sciences* 35:241–272.

Alley, R. B., P. U. Clark, P. Huybrechts, and I. Joughin. 2005. Ice-sheet and sea-level changes. *Science* 310:456–460.

Alley, R. B., J. Marotzke, W. D. Nordhaus, J. T. Overpeck, D. M. Peteet, R. A. Pielke, Jr., R. T. Pierrehumbert, et al. 2003. Abrupt climate change. *Science* 299:2005–2010.

Alley, R. B. 2000. The Younger Dryas cold interval as viewed from central Greenland. *Quaternary Science Reviews* 19:213–226.

Alley, R. B. 2000. *The Two-Mile Time Machine: Ice Cores, Abrupt Climate Change, and Our Future.* Princeton, N.J.: Princeton University Press.

Alley, R. B., D. A. Meese, C. A. Shuman, A. J. Gow, K. C. Taylor, P. M. Grootes, J. W. C. White, et al. 1993. Abrupt increase in snow accumulation at the end of the Younger Dryas event. *Nature* 362:527–529.

Andersen, B. G., and H. W. Borns, Jr. 1994. *The Ice Age World.* Oslo: Scandinavian University Press.

Bard, E., G. Raisbeck, F. Yiou, and J. Jouzeal. 2000. Solar irradiance during the last 1200 years based on cosmogenic nuclides. *Tellus* 528:985–992.

Barlow, L. K., J. C. Rogers, M. C. Serreze, and R. G. Barry. 1997. Aspects of climate variability in the North Atlantic sector: discussion and relation to the Greenland Ice Sheet Project – 2 high resolution isotopic signal. *Journal of Geophysical Research* 102 (C12) (26): 333–344.

Belt, S. T., C. Masse, S. J. Rowland, M. Poulin, and B. LeBlanc. 2007. A novel chemical fossil of palaeo sea ice: IP$_{25}$. *Organic Geochemistry* 38:16–27.

Blytt, A. 1876. *Immigration of the Norwegian Flora. Alb. Cammermeyer.* Oslo: Christiana.

Brauer, A., G. H. Haug, P. Dulski, D. M. Sigman, and J. F. W. Negendank. 2008. An abrupt wind shift in western Europe at the onset of the Younger Dryas cold period. *Nature Geoscience* 1:520–523.

Broecker, W. 2010. *The Great Ocean Conveyor: Discovering the Trigger for Abrupt Climate Change.* Princeton, N.J.: Princeton University Press.

Broecker, W. S., G. H. Denton, R. L. Edwards, H. Cheng, R. B. Alley, and A. E. Putnam. 2010. Putting the Younger Dryas cold event into context. *Quaternary Science Reviews* 29:1078–1081.

Broecker, W. S., and R. Kunzig. 2008. *Fixing Climate –What Past Climate Changes Reveal About the Current Threat and How to Counter It*. New York: Hill and Wang.

Broecker, W. S. 2006. Was the Younger Dryas triggered by a flood? *Science* 312:1146–1148.

Broecker, W. S. 2002. *The Glacial World According to Wally*. Palisades, N.Y.: Eldigio Press, Columbia University, Lamont-Doherty Earth Observatory.

Broecker, W. S. 1998. Paleocean circulation during the last deglaciation: A bipolar seesaw? *Paleoceanography* 13:119–121.

Broecker, W. S. 1997. Thermohaline circulation, the Achilles heel of our climate system: will man-made CO2 upset the current balance? *Science* 278:1582–1588.

Broecker, W. S., and G. H. Denton. 1989. The role of ocean-atmosphere reorganizations in glacial cycles. *Geochimica et Cos*mochimica Acta 53:2465–2501.

Broecker, W. S., D. M. Peteet, and D. Rind. 1985. Does the ocean-atmosphere system have more than one stable mode of operation? *Nature* 315:21–26.

CCSP. 2009. *Past Climate Variability and Change in the Arctic and at High Latitude. A report by the U.S. Climate Change Program and Subcommittee on Global Change Research.* [R. B. Alley, J. Brigham-Grette, G. H. Miller, L. Polyak, and J. W. C. White (coordinating lead authors)] Reston, Va.: U.S. Geological Survey.

Dansgaard, W. 2005. *Frozen Annals: Greenland Ice Cap Research*. Denmark: Niels Bohr Institute, University of Copenhagen.

Denton, G. H., and W. S. Broecker. 2008. Wobbly ocean conveyor circulation during the Holocene? *Quaternary Science Reviews* 27:1939–1950.

Denton, G. H., R. B. Alley, G. C. Comer, and W. S. Broecker. 2005. The role of seasonality in abrupt climate change. Quaternary Science Reviews 24:1159–1182.

Denton, G. H., and T. J. Hughes, eds. 1981. *The Last Great Ice Sheets*. New York: Wiley.

Denton, G. H., and W. Karlen. 1973. Holocene climatic variations—their pattern and possible cause. *Quaternary Research* 3:155–205.

Denton, G. H., and M. Stuiver. 1966. Neoglacial chronology, northeastern St. Elias Mountains, Canada. *American Journal of Science* 264:577–599.

Enfield, D. B., A. M. Mestas-Nuñoz, and P. J. Trimble. 2001. The Atlantic multidecadal oscillation and its relation to rainfall and river flows in the continental U.S. G. H. Denton and S. C. Porter. 1970. Neoglaciation. *Scientific American* 222:100–110.

Fitzhugh, W. W., and E. I. Ward, eds. 2000. *Vikings: The North Atlantic Saga*. Washington, D.C.: Smithsonian Institution Press.

Funder, S. 1989. Quaternary geology of East Greenland. In *Quaternary Geology of Canada and Greenland,* ed. R. J. Fulton. Geological Survey of Canada, pp. 743–822. No. 1 (Chapter 13).

Grove, J. M. 2004. *Little Ice Ages: Ancient and Modern*. 2nd ed. London: Routledge.

Hall, B. L., C. Baroni, and G. H. Denton. 2008. The most extensive Holocene advance in the Stauning Alper, East Greenland, occurred in the Little Ice Age. *Polar Research* 27:128–134.

Herron, E. R., A. Bauder, M. Hoelzle, and M. Maisch eds. 2002. *The Swiss Glaciers 1999/2001 and 2000/2001*. Glaciological Report no. 121/122, 72 pp. Publication of the Glaciological Commission of the Swiss Academy of Sciences.

Holzhauser, H. 1997. Fluctuations of the Grosser Aletsch Glacier and the Gorner Glacier during the past 3200 years; new results. *Paläoklimeforschung Palaeoclimate Research* 24 (Special Issue 16): 35–58.

Holzhauser, H., and H. J. Zumbühl. 1999. Glacier fluctuations in the western Swiss and French Alps in the 16th Century. *Climatic Change* 43:223–237.

Holzhauser, H., M. Magny, and H. J. Zumbühl. 2005. Glacier and lake-level variation in west-central Europe over the last 3500 years. *Holocene* 15:789–801.

Huybers, P., and G. Denton. 2008. Antarctic temperature at orbital timescales controlled by local summer duration. *Nature Geoscience* 1:787–792.

Imbrie, J., and K. P. Imbrie. 1979. *Ice Ages: Solving the Mystery*. Cambridge, Mass.: Harvard University Press.

IPCC. 2007. Summary for Policymakers. In *Climate Change 2007: The Physical Science Basis. Contribution of Working Group I to the Fourth Assessment Report of the Intergovernmental Panel on Climate Change*, ed. S. Solomon, D. Qin, M. Manning, Z. Chen, M. Marquis, K. B. Averyt, M. Tignor, and H. L. Miller. Cambridge: Cambridge University Press.

Jennings, A. E., and N. J. Weiner. 1996. Environmental change in eastern Greenland during the last 1,300 years: Evidence from forminifera and lithofaces in Nansen Fjord, 68 Degrees N. *Holocene* 6 (2):179–191.

Karlen, W., and G. H. Denton. 1975. Holocene glacier variations in Sarek National Park, northern Sweden. *Boreas* 5:25–56.

Kerr, R. A. 2000. A North Atlantic climate pacemaker for the centuries. *Science* 288:1984–1985.

Kerr, R. A. 2005. Atlantic climate pacemaker for millennia past, decades hence? *Science* 309:41–42.

The King's Mirror (Speculum regale). 1917. Trans. Laurence Marcellus Larson. New York: Orchard Books.

Koch, L. 1945. The East Greenland ice. *Meddelelser om Grønland* 130:1–374.

Lamb, H. H. 1979. Climatic variation and changes in the wind and ocean circulation: the Little Ice Age in the northeast Atlantic. *Quaternary Research* 11:1–20.

Lamb, H. H. 1977. *Climate: Present, Past, and Future*. Vol. 2, *Climatic History and the Future*. London: Methuen.

Lamb, H. H. 1965. The early medieval warm epoch and its sequel. *Palaeogeography, Palaeoclimatology, Palaeoecology* 1:12–37.

Lamb, H. H. 1966. *The Changing Climate*. London: Methuen.

Langway, C. C., Jr. 2008. The history of early polar ice cores, US Army Cold Regions Research and Engineering Laboratory ERDC/CRREL TR-08-1.

Lawrence, D. B. 1950. Estimating dates of recent glacier advances and recession rates by studying tree growth layers. *Transactions - American Geophysical Union* 31:243–248.

Lemke, P., J. Ren, R. B. Alley, I. Allison, J. Carrasco, G. Flato, Y. Fujii, et al. 2007. Observations: Changes in Snow, Ice and Frozen Ground. In *Climate Change 2007: The Physical Science Basis. Contribution of Working Group I to the Fourth Assessment Report of the Intergovernmental Panel on Climate Change*, ed. S. Solomon, D. Qin, M. Manning, Z. Chen, M. Marquis, K. B. Averyt, M. Tignor, and H. L. Miller. Cambridge: Cambridge University Press.

Long, A. J. 2009. Back to the future: Greenland's contribution to sea-level change. *GSA Today* 19:4–10.

Luckman, B. H. 2000. The Little Ice Age in the Canadian Rockies. *Geomorphology* 32:357–384.

Lynnerup, N. 1998. The Greenland Norse: A biological-anthropological study. *Meddelelser om Grønland. Man & Society* 24:1–149.

Mangerud, J., S. T. Andersen, B. I. Berglund, and J. J. Donner. 1974. Quaternary stratigraphy of Norden, a proposal for terminology and classification. *Boreas* 3:109–128.

Matthes, F. E., H. F. Reid, W. H. Hobbs, J. E. Church, L. Martin, W. O. Field, E. A. Trager, et al. 1939. Report of committee on glaciers, April 1939. *Transactions - American Geophysical Union* 20:518–523.

Milankovitch, M. 1941. *Canon of Insolation and the Ice Age Problem*. Belgrade: Agency for Textbooks.

National Research Council Committee on Abrupt Climate Change (R.B. Alley, J. Marotzke, W. Nordhaus, J. Overpeck, D. Peteet, R. Pielke, Jr., R. Pierrehumbert, P. Rhines, T. Stocker, L. Talley and J. M. Wallace). 2002. *Abrupt Climate Change: Inevitable Surprises*. Washington, D.C.: National Academy Press.

Ogilvie, A. E. J. 1991. Climatic changes in Iceland c. 865 to 1595. *Acta Archaeologica* 61:233–251.

Ogilvie, A. E. J. 1994. Documentary records of climate from Iceland during the late Mauder Minimum period AD 1675 to 1715, with reference to the isotopic record from Greenland. In *Climatic Trends and Anomalies in Europe 1675-1715*, ed. B. Frenzel. Stuttgart: Fischer Verlag, 9–22.

Reyes, A. V., B. H. Luckman, D. J. Smith, J. J. Clague, and R. D. Van Dorp. 2006. Little Ice Age advance of Kaskawulsh Glacier, St. Elias Mountains, Canada. *Arctic* 59:14–20.

Reyes, A. V., G. C. Wiles, D. J. Smith, D. J. Barclay, S. Allen, S. Jackson, S. Larocque, et al. 2006. Expansion of alpine glaciers in Pacific North America in the first millennium A.D. *Geology* 34:57–60.

Renssen, H., and R. F. B. P. Isarin. 1998. Surface temperature in NW Europe during the Younger Dryas. AGCM simulation compared with temperature reconstructions. *Climate Dynamics* 14:33–44.

Renssen, H., M. Lautenschlager, and C. J. E. Schuurmans. 1966. The atmospheric winter circulation during the Younger Dryas stadial in the Atlantic/European sector. *Climate Dynamics* 12:813–824.

Schlesinger, M. E., and N. Ramankutty. 1994. An oscillation in the global climate system of period 65–70 years. *Nature* 367:723–726.

Seaver, K. 1996. *The Frozen Echo—Greenland and the Exploration of North America ca. A.D. 1000–1500.* Stanford, CA: Stanford University Press.

Serander, R. 1894. *Studier ufer den Gotlandska vegetationens utvecklinghistoria.* Uppsala: Akademisk afthandling.

Sicre, M.-A., J. Jacob, U. Ezat, S. Rousse, C. Kissel, P. Yiou, J. Eiriksson, K. L. Knudsen, E. Jansen, and J.-L. Turon. 2008. Decadal variability of sea-surface temperature off North Iceland over the last 2000 years. *Earth and Planetary Science Letters* 268:142–147.

Sutton, R. T., and D. L. R. Hodson. 2005. Atlantic Ocean forcing of North American and European summer climate. *Science* 309:115–118.

Sutton, R. T., and D. L. R. Hodson. 2003. Influence of the ocean on North Atlantic climate variability 1871–1999. *Journal of Climate* 16:3296–3313.

Taylor, K. C., G. W. Lamorey, G. A. Doyle, R. B. Alley, P. M. Grootes, P. A. Mayewski, J. W. C. White, and L. K. Barlow. 1993. The 'flickering switch' of late Pleistocene climate change. *Nature* 361:432–436.

Von Post, L. 1967. Forest tree pollen in South Swedish peat bog deposits. *Pollen et Spores* 9:375–401.

Wang, Y. J., H. Cheng, R. L. Edwards, Y. Q. He, X. G. Kong, Z. S. An, J. Y. Wu, M. J. Kelly, C. A. Dykoski, and X. D. Li. 2005. The Holocene Asian monsoon: Links to solar changes and North Atlantic climate. *Science* 308:854–857.

Wagner, B., and M. Melles. 2002. Holocene environmental history of Ymen Ø, East Greenland, inferred from lake sediments. *Quaternary International* 89:165–176.

Weidick, A., M. Kelly, and O. Bennike. 2004. Late Quaternary development of the southern sector of the Greenland Ice Sheet, with particular reference to the Qassimiut lobe. *Boreas* 33:284–299.

Wiles, G. C., D. J. Barclay, P. E. Calkin, and T. V. Lowell. 2008. Century to millennial-scale temperature variations for the last 2000 years inferred from glacial geologic records of southern Alaska. *Global and Planetary Change* 60:115–125.

Zumbühl, H. J., D. Steiner, and S. V. Nussbaumer. 2007. 19th century glacier representations and fluctuations in the central and western European Alps: An interdisciplinary approach. *Global and Planetary Change* 60:42–57.

INDEX